18

CRM
SERIES

Centro
di Ricerca
Matematica
Ennio De Giorgi

Geometry, Structure and Randomness in Combinatorics

edited by
Jiří Matoušek, Jaroslav Nešetřil
and Marco Pellegrini

EDIZIONI
DELLA
NORMALE

© 2015 Scuola Normale Superiore Pisa

ISBN 978-88-7642-524-0
e-ISBN 978-88-7642-525-7

Contents

Preface

On September 3-7, 2012, as part of the activities of the Mathematics Research Center "Ennio De Giorgi" and on the invitation of its director prof. Mariano Giaquinta, we organized the Workshop "Geometry, Structure and Randomness in Combinatorics" at Scuola Normale Superiore in Pisa. The workshop was organized by Jiří Matousek, Jaroslav Nešetřil (Charles University, Prague) and Marco Pellegrini (CNR, Pisa) and has been supported jointly by SNS and CRM Pisa and DIMATIA centre in Prague.

This workshop intended to reflect some key recent advances in combinatorics, particularly in the area of extremal theory and Ramsey theory. It also aimed to demonstrate the broad spectrum of techniques and its relationship to other fields of mathematics, particularly to geometry, logic and number theory.

Invited speakers included ten of the leading experts. We had the pleasure to invite Prof. Endre Szemerédi, the winner of the Abel Prize in 2012 for his fundamental contributions in the field of discrete mathematics and theoretical computer science. The workshop attracted 48 participants both from Italy and abroad.

The following list is that of the invited lectures at the workshop:

IMRE BÁRÁNY, *Tensors, colours, and octahedral*

BÉLA BOLLOBÁS, *Extremal and probabilistic results on bootstrap percolation*

MARIA CHUDNOVSKY, *Extending the Gyarfas-Sumner conjecture*

ZEEV DVIR, *Configurations of points with many collinear triples: going beyond Sylvester-Gallai*

ZOLTÁN FÜREDI, *Binary codes versus hypergraphs*

JAROSLAV NEŠETŘIL, *A unifying approach to graph limits II*

PATRICE OSSONA DE MENDEZ, *A unifying approach to graph limits I*

ALEX SCOTT, *Discrepancy in graphs, hypergraphs and tournaments*
and (second talk)
Szemerédi regularity lemma for sparse graphs

JÓZSEF SOLYMOSI, *Sums vs. products*
and (second talk)
The (7,4)-conjecture for finite groups

ENDRE SZEMERÉDI, *On subset sums*

Given the success both scientific and public of the workshops, at the end
of the event, at the suggestion of Professor Mariano Giaquinta, it has been
proposed to organize a volume dedicated to this meeting. This proposal
was welcomed by all the speakers. The present volume has been edited
for the "CRM Series", with the title "Geometry, Structure and Randomness in Combinatorics" and includes both original scientific articles in
extended form or survey articles on results and problems inherent in the
themes presented at the workshop. Each article submitted was reviewed.

We thank all the authors for their contribution and again Scuola Normale Superiore and its Centro di Ricerca Matematica Ennio De Giorgi
and to DIMATIA Centre of Charles University for their generous support.

Pisa/Prague

Jiří Matoušek, Jaroslav Nešetřil, Marco Pellegrini

Authors' affiliations

IMRE BÁRÁNY – Alfréd Rényi Institute of Mathematics, Hungarian Academy of Sciences, PO Box 127, 1364 Budapest, Hungary
and
Department of Mathematics, University College London, Gower Street, London WC1E 6BT, UK

BÉLA BOLLOBÁS – Department of Pure Mathematics and Mathematical Statistics, Wilberforce Road, Cambridge CB3 0WB, UK
and
Department of Mathematical Sciences, University of Memphis, Memphis TN 38152, USA

MARIA CHUDNOVSKY – Department of Mathematics, Columbia University, 308 Mudd Bldg, 2990 Broadway, New York NY 10027, USA

MAURO DI NASSO – Dipartimento di Matematica, Università di Pisa, Largo Bruno Pontecorvo 5, 56127 Pisa, Italia

ZOLTÁN FÜREDI – Alfréd Rényi Institute of Mathematics, Hungarian Academy of Sciences, 13–15 Reáltanoda Street, 1053 Budapest, Hungary

PETER HEGARTY – Department of Mathematical Sciences, University of Gothenburg, Chalmers Tvärgata 3, 41261 Göteborg, Sweden

IDA KANTOR – Computer Science Institute of Charles University, Malostranské nám. 25, 118 00 Praha 1, Czech Republic

GYULA O. H. KATONA – Alfréd Rényi Institute of Mathematics, Hungarian Academy of Sciences, Reáltanoda u. 13–15, Budapest 1053, Hungary

IMRE LEADER – Department of Pure Mathematics and Mathematical Statistics, Wilberforce Road, Cambridge CB3 0WB, UK

JIŘÍ MATOUŠEK – Department of Applied Mathematics, Charles University, Malostranské nám. 25, 118 00 Praha 1, Czech Republic
and
Institute of Theoretical Computer Science, ETH Zürich, 8092 Zürich, Switzerland

JAROSLAV NEŠETŘIL – Department of Applied Mathematics, Charles University and Institute for Theoretical Computer Science (ITI), Malostranské nám.25, 11800 Praha 1, Czech Republic

PATRICE OSSONA DE MENDEZ – Centre d'Analyse et de Mathématiques Sociales (CNRS, UMR 8557), 190-198 avenue de France, 75013 Paris, France

RYAN SCHWARTZ – Department of Mathematics, University of British Columbia, Vancouver, B.C., V6T1Z2, Canada

JÓZSEF SOLYMOSI – Department of Mathematics, University of British Columbia, Vancouver, B.C., V6T1Z2, Canada

DMITRY ZHELEZOV – Department of Mathematical Sciences, University of Gothenburg, Chalmers Tvärgata 3, 41261 Göteborg, Sweden

Tensors, colours, octahedra

Imre Bárány

Abstract. Several theorems in combinatorial convexity admit colourful versions. This survey describes old and new applications of two methods that can give such colourful results. One is the octahedral construction, the other is Sarkaria's tensor method.

1 Introduction

Theorems of Carathéodory, Helly, and Tverberg are classical results in combinatorial convexity. They all have coloured versions. Some others involve colours directly. For instance in Kirchberger's theorem [15], the elements of a finite set $X \subset \mathbb{R}^d$ are coloured Red and Blue, and the statement is that the Red and Blue points can be separated by a hyperplane if and only if for every $Y \subset X$ with $|Y| \leq d + 2$, the Red and Blue points in Y can be separated by a hyperplane.

The aim of this paper is to describe and explain old and new applications of two methods that have turned out to be useful when proving such colourful results. One is the octahedral construction, discovered and first used by László Lovász in 1991, which appeared in [4]. The other is Karinbir Sarkaria's tensor method, originally from [25] and developed further in [5].

In the next section Tverberg's theorem and its colourful version are presented. The octahedral construction is given in Section 3 with applications followed in later sections.

2 Tverberg's theorem and its coloured version

Tverberg's theorem is a gem, one of my favourites. Here is what it says.

Theorem 2.1. *Assume $d \geq 1, r \geq 2$ and $X \subset \mathbb{R}^d$ has $(r - 1)(d + 1) + 1$ elements. Then X has a partition into r parts $X_1, \ldots, X_r$ such that $\bigcap_1^r \operatorname{conv} X_i \neq \emptyset$.*

The number $(r-1)(d+1)+1$ is best possible here: for a general position X with one fewer element, the affine hulls of an r-partition do not have a common point (by counting dimensions).

The case $r = 2$ is Radon's theorem from 1922 [21] that has a simple proof: Given $x \in \mathbb{R}^d$ we write $(x, 1)$ for the $(d+1)$-dimensional vector whose first d components are equal to those of x, and the last one is 1. This time $|X| = d+2$ so the vectors $(x, 1) \in \mathbb{R}^{d+1}$ have a nontrivial linear dependence $\sum \alpha(x)(x, 1) = (0, 0)$. Letting $X_1 = \{x \in X : \alpha(x) \geq 0\}$ and $X_2 = \{x \in X : \alpha(x) < 0\}$ is the partition needed. Indeed, defining $\alpha = \sum_{x \in X_1} \alpha(x)$ and $\alpha^*(x) = \alpha(x)/\alpha$ for $x \in X_1$ and $\alpha^*(x) = -\alpha(x)/\alpha$ for $x \in X_2$, we have convex combinations in

$$z = \sum_{x \in X_1} \alpha^*(x)x = \sum_{x \in X_2} \alpha^*(x)x$$

showing that $z \in \operatorname{conv} X_1 \bigcap \operatorname{conv} X_2$.

There are several proofs of Tverberg's theorem, for instance in Tverberg [29, 30], Tverberg and Vrećica [31], Roudneff [23], Sarkaria [25], Bárány and Onn [5], Zvagelskii [34], none of them easy. We will give another proof in Section 8 which is from Arocha *et al.* [1].

The coloured version of Tverberg's theorem follows now.

Theorem 2.2. *For every $d \geq 1$ and $r \geq 2$ there is $t = t(r, d)$ with the following property. Given sets $C_1, \ldots, C_{d+1} \in \mathbb{R}^d$ (called colours), each of size t, there are r disjoint sets $S_1, \ldots, S_r \subset \bigcup_1^{d+1} C_i$ such that $|S_j \cap C_i| = 1$ for every i, j and $\bigcap_1^r \operatorname{conv} S_j \neq \emptyset$.*

In other words, given colours $C_1, \ldots, C_{d+1} \subset \mathbb{R}^d$ of large enough size, there are r disjoint and colourful sets S_j whose convex hulls have a point in common. Colourful means that S_j is a transversal of the C_i, that is, S_j contains one element from each C_i. The need for this result emerged in connection with the halving plane problem (*c.f.* [3]). It was proved there that $t(3, 2)$ is finite. Shortly afterward it was proved by Bárány and Larman [4] that $t(r, 2) = r$ for all r, clearly the best possible result. The same paper presents Lovász's proof that $t(2, d) = 2$ for all d, the first application of the octahedral method. To simplify notation we write $[k]$ for the set $\{1, 2, \ldots, k\}$.

3   The octahedral construction

Proof of $t(2, d) = 2$. We have $C_i = \{a_i, b_i\} \subset \mathbb{R}^d$, $i \in [d+1]$. Note that we may exchange the names of a_i and b_i later. We want to choose a transversal T from $C_1, \ldots, C_{d+1}$ such that the convex hulls of T and of the complementary transversal $\overline{T}$ have a point in common. For this purpose let

$$Q^{d+1} = \text{conv}\{\pm e_1, \ldots, \pm e_{d+1}\}$$

be the standard octahedron in $\mathbb{R}^{d+1}$ (the e_i are the usual basis vectors). We define a map $f : \partial Q^{d+1} \to \mathbb{R}^d$ by setting $f(e_i) = a_i$ and $f(-e_i) = b_i$, and then extend f simplicially to ∂Q^{d+1}, that is, to the facets of Q^{d+1}. Note that f maps the facets of Q^{d+1} to the convex hull of a transversal T exactly, and the opposite facet is mapped to conv $\overline{T}$. So what we need is a pair of opposite facets whose images intersect.

This cries out for the Borsuk-Ulam theorem: ∂Q^{d+1} is homeomorphic to S^d and so f is an $S^d \to \mathbb{R}^d$ map. By a variant of Borsuk-Ulam there are antipodal points $z, -z \in \partial Q^{d+1}$ with $f(z) = f(-z)$. If z lies on a facet F, then $-z$ lies on the opposite facet $\overline{F}$. For simpler writing assume that $F = \text{conv}\{e_1, \ldots, e_{d+1}\}$, then $\overline{F} = \text{conv}\{-e_1, \ldots, -e_{d+1}\}$, and we see that $\text{conv}\{a_1, \ldots, a_{d+1}\}$ and $\text{conv}\{b_1, \ldots, b_{d+1}\}$ have $f(z) = f(-z)$ as a common point.

Actually, more is true: if $z = \sum_1^{d+1} \gamma_i e_i$, then $-z = \sum_1^{d+1} \gamma_i(-e_i)$, and the common point is $\sum_1^{d+1} \gamma_i a_i = \sum_1^{d+1} \gamma_i b_i$. Thus the common point comes with the same coefficients in the convex combinations. $\square$

This is the octahedral method. The basic idea is that facets of the octahedron correspond to transversals of $C_1, \ldots, C_{d+1}$, transversals have the structure of ∂Q^{d+1}, and disjoint transversals come from opposite facets, and the next step is the use of algebraic topology like the Borsuk-Ulam theorem above.

Unfortunately the method does not work for $r \geq 3$. It was conjectured in [4] that $t(r, d) = r$ for all r and d. Finiteness of $t(r, d)$ was proved by Živaljević and Vrećica [33] using equivariant topology. Their result is that $t(r, d) \leq 2r - 1$ if r is a prime (which implies finiteness of $t(r, d)$ for all r). The same was proved by different methods by Björner et al. [8] and by Matoušek [17]. More recently Blagojević, Matschke, and Ziegler [9] showed that $t(r, d) = r$ if $r + 1$ is a prime which is again best possible. The strange primality condition in all cases is needed because cyclic groups of prime order behave better in equivariant topology. But the theorem is probably true for every r, the primality condition is required for the method and not for the problem. It is however disappointing (for a convex geometer) that a completely convex (or linear, if you wish) problem does not have a direct convex (or linear) proof, and topology seems a necessity here. Finding a purely geometric proof remains a challenge. The interested reader may wish to read Günter Ziegler's fascinating article [32] about Tverberg's theorem and its colourful version.

Remark 3.1. Quite recently, Pablo Soberón [26] has found another (and simpler) proof of $t(2, d) = 2$. It starts with the observation that the vectors $a_i - b_i$, $i \in [d + 1]$ are linearly dependent, so $\sum_1^{d+1} \gamma_i (a_i - b_i) = 0$ for some not all zero γ_i. Some γ_i may be negative, but then we exchange the names of a_i and b_i which makes γ_i positive. Then $\sum \gamma_i a_i = \sum \gamma_i b_i$. Assuming as we can that $\sum \gamma_i = 1$, $\sum \gamma_i a_i = \sum \gamma_i b_i$ is a common point of the convex hulls of transversals $a_1, \ldots, a_{d+1}$ and $b_1, \ldots, b_{d+1}$. Note that even the coefficients are the same. So this method gives exactly the same result as the octahedral construction. A little extra is the efficient algorithm that follows from this proof. The paper [26] gives precise conditions for the existence of colourful partitions whose convex hulls have a common point with equal coefficients. The proof uses tensors as in Sarkaria's lemma which will be described in Section 7.

4   Colourful Carathéodory theorem

Carathéodory's classical theorem says in essence that being in the convex hull has a very finite reason. Precisely, if $A \subset \mathbb{R}^d$ and $a \in \text{conv } A$, then $a \in \text{conv } B$ for some $B \subset A$ with $|B| \leq d + 1$. The colourful version of this theorem is an old result of mine [2].

Theorem 4.1. *If $A_1, \ldots, A_{d+1} \subset \mathbb{R}^d$ and $a \in \bigcap_1^{d+1} \text{conv } A_i$, then there is a transversal $a_i \in A_i$ for all i, such that $a \in \text{conv}\{a_1, \ldots, a_{d+1}\}$.*

The colourful version contains the original one: simply take $A_i = A$ for every A. A natural question is how many such transversals are there, and the natural setting for the question is when a is the origin (which makes no difference), the points in $\bigcup_1^{d+1} A_i$ together with the origin are in general position, and each A_i has exactly $d + 1$ elements. Of course, $0 \notin \bigcup_1^{d+1} A_i$, and we may assume that each $A_i \subset S^{d-1}$, the unit sphere of $\mathbb{R}^d$. We call a transversal $\{a_1, \ldots, a_{d+1}\}$ special if the origin lies in its convex hull. Define $\tau(d)$ as the minimal number of special transversals under these conditions.

A neat construction from Deza et al. [10] shows that $\tau(d) \leq d^2 + 1$ and it is not hard to check that $\tau(2) = 5$. Carathéodory's theorem has a cone hull or positive hull version, slightly stronger than the convex one. We write $\text{pos } A$ for the cone hull of the elements in $A \subset \mathbb{R}^d$, that is, $\text{pos } A$ is the set of vectors $\sum_1^n \gamma_i a_i$ with $\gamma_i \geq 0$ and $a_i \in A$ for all $i \in [n]$ and all $n \in \mathbb{Z}$.

Theorem 4.2. *If $A_1, \ldots, A_d \subset \mathbb{R}^d$ and $a \in \bigcap_1^d \text{pos } A_i$, then there is a transversal $a_i \in A_i$ for all i such that $a \in \text{pos}\{a_1, \ldots, a_d\}$.*

It is quite easy to check (or see [2] for the proof) that this result has the following consequence.

Corollary 4.3. *If $a \in \mathbb{R}^d$, $A_1, \ldots, A_d \subset \mathbb{R}^d$, and $0 \in \bigcup_1^d \text{conv } A_i$, then there is a transversal $a_i \in A_i$ for all i such that $0 \in \text{conv}\{a, a_1, \ldots, a_d\}$.*

The Corollary shows immediately that every point in $\bigcup_1^{d+1} A_i$ belongs to at least one special transversal, so $\tau(d) \geq d + 1$. The octahedral construction has been used to improve this bound to a quadratic one, in several papers. In particular, Bárány and Matoušek [6] show $\tau(d) \geq d(d + 1)/5$ and $\tau(3) = 10$ (which is best possible), Stephen and Thomas [28] prove $\tau(d) \geq (d + 2)^2/4$, and Deza *et al.* [11] give $\tau(d) \geq (d + 1)^2/2$, which is further improved to $\tau(d) \geq \frac{1}{2}d^2 + \frac{7}{2}d - 8$ when $d \geq 4$ by Deza, Meunier, and Sarrabezolles in [12].

How can the octahedral construction help here? Well, it is clear that $a_1, \ldots, a_{d+1}$ is a special transversal iff $-a_{d+1} \in \text{pos}\{a_1, \ldots, a_d\}$. Fix now the special transversal $a_1, \ldots, a_{d+1}$ and consider a partial transversal $b_1 \in A_1, \ldots, b_d \in A_d$ with b_i different from a_i for all i. The octahedral construction defines a map $f : \partial Q^d \to \mathbb{R}^d$ by setting $f(e_i) = a_i$, $f(-e_i) = b_i$ (for all i), and then extend it simplicially to ∂Q^d. As $0 \notin f(\partial Q^d)$ by the general position assumption, we can define $g(x) = f(x)/\|f(x)\|$. The map $g : \partial Q^d \to S^{d-1}$ is continuous and is essentially an $S^{d-1} \to S^{d-1}$ map, so if it takes some (non-critical) value, then it takes it at least twice, or else it takes every value at least once.

More precisely, if the degree of g is nonzero, then g takes every value in S^{d-1} at least once, and if its degree is zero, then it takes every non-critical (in the sense of Sard's Lemma, see Milnor's book [18]) value at least twice. But g takes the value $-a_{d+1}$ at least once, since $-a_{d+1} \in g(\partial Q^d)$. Moreover, this value is non-critical because of the general position assumption. Writing $T = \{a_1, \ldots, a_{d+1}\}$ and $B = \{b_1, \ldots, b_d\}$ we have established the following fact.

Lemma 4.4. *Under the above condition either $T \cup B$ contains another special transversal, different from T, or every $b_{d+1} \in A_{d+1} \setminus \{a_{d+1}\}$ belongs to a special transversal from $T \cup B$.*

This consequence of the octahedral construction is used, with varying outcome, in all quadratic lower bounds to $\tau(d)$. But the lemma also leads to a completely combinatorial problem: determine the minimum number of edges a hypergraph $\mathcal{H}$ can have provided it is a $(d+1)$-partite $(d+1)$-uniform hypergraph with partition classes $A_1, \ldots, A_{d+1}$, $|A_i| = d + 1$ for each i, and satisfies the following conditions (mimicking those of the special transversals):

- for every $a \in \bigcup_1^{d+1} A_i$ there is $T \in \mathcal{H}$ with $a \in T$
- for every i and for every $T \in \mathcal{H}$ with $T \cap A_i = a_i$, and for every $B = \{b_1, \ldots, b_{i-1}, b_{i+1}, \ldots, b_{d+1}\}$ with B disjoint from T, either there is $T^* \in \mathcal{H}$, $T^* \neq T$ with $T^* \subset T \cup B$, or for every $a \in A_i$ there is $T^* \in \mathcal{H}$ with $a \in T^*$ and $T^* \setminus \{a\} \subset T \cup B$.

Here the first condition comes from Corollary 4.3, and the second from Lemma 4.4 as the role of a_{d+1} can be taken be an arbitrary $a \in \bigcup A_i$. Note however that the condition $0 \in \text{conv } A_i$ is lost in this combinatorial setting.

Open question 4.5. For a hypergraph $\mathcal{H}$ with these properties, does $|\mathcal{H}|$ have to have at least $d^2 + 1$ edges? [1] Even with no hypergraph, is it true that $\tau(d) = d^2 + 1$?

5   Colourful Carathéodory strengthened

The following result is a generalization of Theorem 4.1. It was found at the same time on two different continents, and was published by Holmsen, Pach, Tverberg [14] and by Arocha, Bárány, Fabila, Bracho, Montejano [1]. The proof is based on the octahedral construction. In both cases the original target was a colourful Helly type theorem on the sphere, see [14] or [1].

Theorem 5.1. *If $A_1, \ldots, A_{d+1} \subset \mathbb{R}^d$, none of them empty and $a \in \text{conv}(A_i \cup A_j)$ for all distinct $i, j \in [d+1]$, then there is a transversal $a_i \in A_i$ for all i, such that $a \in \text{conv}\{a_1, \ldots, a_{d+1}\}$.*

Proof. We identify a with the origin. We can assume that every A_i is finite. Let T be the transversal with $a_i \in A_i$ for $i \in [d+1]$ whose convex hull $\triangle = \text{conv } T$ is closest to the origin. Let $z \in \triangle$ be this closest point. If $z = 0$ we are done, so assume $z \neq 0$, and let H be the hyperplane passing through, and orthogonal to, z. Write H^+ for the closed halfspace bounded by H and not containing 0. As z is on the boundary of the simplex $\triangle$, it is in the convex hull of a proper subset of T, say of $\{a_1, \ldots, a_d\}$.

We claim that z lies in the relative interior of $\text{conv}\{a_1, \ldots, a_d\}$. Assume on the contrary that z is in the convex hull of $\{a_1, \ldots, a_{d-1}\}$, say. There is a point $b \in (A_d \cup A_{d+1}) \setminus H^+$ as otherwise $A_d \cup A_{d+1} \subset H^+$ so their convex hull does not contain the origin, contrary to the condition of

---

[1] This question has recently been settled in the affirmative by Pauline Sarrabezolles [24], implying $\tau(d) = d^2 + 1$.

the theorem. Now $\{a_1, \ldots, a_{d-1}, b\}$ can be extended to a transversal T^* whose convex hull of contains the segment $[z, b]$. But $[z, b]$, and consequently conv T^*, contains a point closer to the origin than z, contradicting the choice of T.

Note that $A_{d+1} \subset H^+$ since replacing a_{d+1} by any $b \in A_{d+1} \setminus H^+$ would give a transversal whose convex hull is closer to the origin than conv T. Let H_0 be the hyperplane parallel to H and containing the origin, and H^- be the closed halfspace bounded by H_0 and not containing $\triangle$.

It follows that there is $b_i \in A_i \cap H^-$ for every $i \in [d]$, as otherwise $A_i \cup A_{d+1}$ lies in the complement of H^- and is then separated from the origin. We can apply the octahedral construction now. Define $f : \partial Q^d \to \mathbb{R}^d$ by setting $f(e_i) = a_i$, $f(-e_i) = b_i$ and extend f simplicially to the facets of Q^d.

Again, ∂Q^d is an S^{d-1} so removing $f(\partial Q^d)$ from $\mathbb{R}^d$ results in an unbounded connected component and finitely many bounded connected components (by the Jordan curve theorem in higher dimensions). The unbounded component contains the interior of H^+. The segment $[0, z)$ is disjoint from $f(\partial Q^d)$ so it lies in some connected component Ω. It is clear that Ω is not the unbounded connected component.

Consider now a point $a \in A_{d+1}$, and the halfline L starting at 0 in direction $-a$. L starts in Ω and ends up in the unbounded component. So it must intersect $f(\partial Q^d)$ at some point $v = L \cap f(F)$ where F is a facet of ∂Q^d. Then v is in the convex hull of a transversal of $A_1, \ldots, A_d$ (even of $\{a_1, b_1\}, \ldots, \{a_{d+1}, b_{d+1}\}$). Since $0 \in [a, v]$, the convex hull of this partial transversal and a contains the origin, contrary to the choice of T. $\square$

The theorem has $d(d+1)/2$ conditions, one for each pair i, j. All of them are needed as the following example shows. Assume the points $a, x_1, \ldots, x_{d+1}, y$ are in general position and $a \in \text{conv}\{x_1, \ldots, x_{d+1}\}$, and let $A_i = \{x_1, \ldots, x_{d+1}\}$ for $i \in [d-1]$ and $A_d = A_{d+1} = \{y\}$. There is no transversal whose convex hull would contain a yet for every pair i, j apart from $d, d+1$, $a \in \text{conv}(A_i \cup A_j)$. The same example shows that the conditions $a \in \text{conv}(A_i \cup A_j \cup A_k)$ for every triple i, j, k do not work. More disappointingly, the result does not extend to the cone hull, as shown by a very simple example in $\mathbb{R}^2$.

Open question 5.2. It would be interesting to design an effective algorithm that, under the conditions of Theorem 4.1, finds a colourful simplex whose convex hull contains the origin. The proof of Theorem 4.1, and also that of Theorem 5.1 only gives the existence of such a simplex. So in fixed dimension they give an algorithm with at most $(d+1)^{d+1}$ steps,

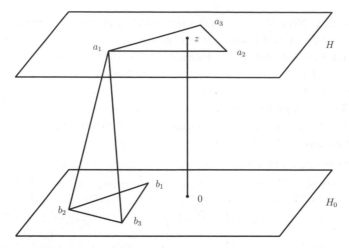

Figure 5.1. Figure for Theorem 6.1, almost works for Theorem 5.1. Only parts of $f(\partial Q^d)$ are drawn.

which is way too many when dimension is part of the input. For further information about this question see Bárány, Onn [5].

6   Colourful Carathéodory for connected compacta

A theorem of Fenchel [13] asserts that if a compact set $A \subset \mathbb{R}^d$ is connected, then $a \in \operatorname{conv} A$ implies the existence of $B \subset A$ with $a \in \operatorname{conv} B$ and $|B| \leq d$. So the Carathéodory number can be lowered. The colourful version of Fenchel's result is given in Bárány and Karasev [7]. Its proof is based on the octahedral construction, again. Recall that if $A \subset \mathbb{R}^d$ is connected and compact, then for every hyperplane H disjoint from A, one of the open halfspaces bounded by H contains A.

Theorem 6.1. *Assume $A_1, \ldots, A_d \subset \mathbb{R}^d$ are compact connected sets with $0 \in \bigcap_1^d \operatorname{conv} A_i$. Then there is a transversal $a_i \in A_i$ ($i \in [d]$ such that $0 \in \operatorname{conv}\{a_1, \ldots, a_d\}$.*

The proof is similar to the previous one. Choose a transversal $T = \{a_1, \ldots, a_d\}$ whose convex hull $\triangle$ is closest to the origin, and let $z \in \triangle$ be this closest point. If $z = 0$ we are done, so suppose $z \neq 0$. It is easy to see (we omit the details) that $\triangle$ is a $d - 1$-dimensional simplex and that z lies in the relative interior of S.

Let H be the hyperplane passing through, and orthogonal to, z, clearly $\triangle \subset H$. Again, let H_0 be the hyperplane parallel to H and containing the origin. As A_i is connected, there is a point $b_i \in H_0 \cap A_i$ for every $i \in [d]$. The octahedral construction applies the same way as before.

So we have $f : \partial Q^d \to \mathbb{R}^d$, the simplicial extension from the vertices $f(e_i) = a_i$, $f(-e_i) = b_i$. This time $f(\partial Q^d)$ lies between H and H_0, and evidently $0 \notin f(\partial Q^d)$. Again, removal of $f(\partial Q^d)$ from $\mathbb{R}^d$ yields connected components, and 0 is in the unbounded one. But the points on the segment $[0, z)$, close enough to z lie in a bounded component. This shows that the open segment $(0, z)$ intersects $f(\partial Q^d)$. The intersection point is in the convex hull of a transversal and closer to the origin than z. Contradiction. □

In [7] a second (and interesting) proof of the theorem is given which uses vector bundles and has some further consequences.

7 Sarkaria's lemma

Assume $X_1, X_2, \ldots, X_r \subset \mathbb{R}^d$ are finite sets, $r \geq 2$. There is a good necessary and sufficient condition for $\bigcap_1^r \operatorname{conv} X_i = \emptyset$ which we now describe.

Theorem 7.1. *Under the above conditions, $\bigcap_1^r \operatorname{conv} X_i = \emptyset$ if and only if there are closed halfspaces $D_1, \ldots, D_r$ with $\operatorname{conv} X_i \subset D_i$ for every $i \in [r]$ such that $\bigcap_1^r D_i = \emptyset$.*

The **proof** is easy. One direction is trivial. For the other one set $K_i = \operatorname{conv} X_i$. The case $r = 2$ is just the separation theorem for convex sets. For larger r we have $K_1 \cap \bigcap_2^r K_i = \emptyset$ so by separation there is a closed halfspace D_1 containing K_1 with $D_1 \cap \bigcap_2^r K_i = \emptyset$. This way K_1 is replaced by D_1, and the same way K_2 is replaced by D_2, etc. After step $j-1$ we have $\bigcap_1^{j-1} D_i \cap \bigcap_j^r K_i = \emptyset$ and so $K_j \cap \left(\bigcap_1^{j-1} D_i \cap \bigcap_{j+1}^r K_i \right) = \emptyset$. Here K_j is convex, compact and $\bigcap_1^{j-1} D_i \cap \bigcap_{j+1}^r K_i$ is convex so the separation theorem applies. □

The case when $\bigcap_1^r D_i = \emptyset$ can be characterized by duality. Assume $D_i = \{x \in \mathbb{R}^d : a_i x \leq \alpha_i\}$.

Theorem 7.2. *With the above notation $\bigcap_1^r D_i = \emptyset$ if and only if $(0, -1) \in \operatorname{pos}\{(a_i, \alpha_i) : i \in [r]\}$.*

Sketch of **proof**. The condition $\bigcap_1^r D_i = \emptyset$ is equivalent to "the system of linear inequalities $a_i x \leq \alpha_i$, $i \in [r]$ has no solution". Then Farkas's lemma proves the theorem. □

This an outer or dual type characterization. Sarkaria's lemma is an inner characterization of the fact that $\bigcap_1^k \operatorname{conv} X_i = \emptyset$. We need an artificial tool: choose vectors $v_1, \ldots, v_r \in \mathbb{R}^{r-1}$ so that their unique (up to a multiplier) linear dependence is $v_1 + \cdots + v_r = 0$. Suppose that

$X_1, X_2, \ldots, X_r \subset \mathbb{R}^d$ are finite sets, and write $X = \bigcup_1^r X_i$. We assume further that the X_is are disjoint. (Alternatively, we can consider X a multiset.) Each $x \in X$ comes from a unique $X_i = X_{i(x)}$. With each such x we associate the tensor

$$\overline{x} = v_i \otimes (x, 1) \in \mathbb{R}^n$$

where $n = (r - 1)(d + 1)$. The tensor $\overline{x}$ can be thought of as an $(r - 1)$ by $(d + 1)$ matrix as well. Here is Sarkaria's lemma [25], in the form given in Bárány and Onn [5]. (Originally it used number fields instead of tensors.)

Theorem 7.3. *Under the above conditions, $\bigcap_1^r \operatorname{conv} X_i = \emptyset$ if and only if $0 \notin \operatorname{conv} \overline{X}$.*

Proof. We prove the theorem by showing that $0 \in \operatorname{conv} \overline{X}$ iff $\bigcap_1^r \operatorname{conv} X_i \neq \emptyset$.

If $0 \in \operatorname{conv} \overline{X}$, then there are $\alpha(x) \geq 0$ for all $x \in X$ such that

$$\sum_{x \in X} \alpha(x) = 1 \text{ and } \sum_{x \in X} \alpha(x)\overline{x} = 0.$$

Replacing $\overline{x}$ by $v_i \otimes (x, 1)$ gives

$$0 = \sum_{x \in X} \alpha(x)\overline{x} = \sum_{i=1}^r \sum_{x \in X_i} \alpha(x)v_i \otimes (x, 1)$$

$$= \sum_{i=1}^r v_i \otimes \sum_{x \in X_i} \alpha(x)(x, 1).$$

Set $z_i = \sum_{x \in X_i} \alpha(x)(x, 1) \in \mathbb{R}^{d+1}$ for $i \in [r]$. We claim that $z_1 = z_2 = \cdots = z_r$. By symmetry it suffices to show that $z_1 = z_2$. By the choice of the vectors $v_1, \ldots, v_r$ there is $u \in \mathbb{R}^{r-1}$ such that $uv_1 = 1$, $uv_2 = -1$ and $uv_i = 0$ for all $i > 2$. Multiplying the last formula by u from the left gives $0 = \sum_{i=1}^r uv_i \otimes z_i = z_1 - z_2$.

This implies, in particular, that the last coordinate of each z_i is equal to $1/r$. Thus $y_i = \sum_{x \in X_i} r\alpha(x)x$ is a convex combination of the elements of X_i, and $y = y_1 = \cdots = y_r$. Consequently y is a common element of each $\operatorname{conv} X_i$.

The steps of this proof can be reversed easily showing that condition $\bigcap_1^r \operatorname{conv} X_i \neq \emptyset$ implies $0 \in \operatorname{conv} \overline{X}$. $\square$

Remark 7.4. Note that when $r = 2$, $v_1 = 1$ and $v_2 = -1$, Sarkaria's lemma gives X_1 respectively X_2 as the set of elements with positive (and negative) coefficients in the linear dependence of $(x_1, 1), \ldots, (x_{d+2}, 1)$. Sarkaria's tensor method is a direct and beautiful generalization of the proof of Radon's theorem.

8   Kirchberger generalized

Recall Kirchberger's theorem [15] from the first section with Red and Blue points. We want to have more colours this time, so we give the theorem in a slightly different form. The elements of a finite sets $X \subset \mathbb{R}^d$ are coloured Red or Blue which is simply a partition of X into X_1, (the Red points) and X_2, (the Blue ones). We say that X is separated along the colours if $\operatorname{conv} X_1 \cap \operatorname{conv} X_2 = \emptyset$. Now Kirchberger's theorem says that X is separated along the colours iff every subset of X, of size at most $d + 2$, is separated along the colours.

The extension to more colours is quite natural now. Assume a finite set (or multiset) $X \subset \mathbb{R}^d$ is coloured with $r \geq 2$ colours, that is, there is a partition $X = X_1 \cup \cdots \cup X_r$. We say that X is separated along the colours if $\bigcap_1^r \operatorname{conv} X_i = \emptyset$. The colourful version of Kirchberger's theorem is a result of A. Pór [20]:

Theorem 8.1. *With the above notation $X \subset \mathbb{R}^d$ is separated along the colours if and only if every subset of X of size at least $(r - 1)(d + 1) + 1$ is separated along the colours.*

Note that $r = 2$ is the original Kirchberger theorem. Theorem 8.1 can be proved using Theorem 7.2, for instance. But here we aim for more. Set $n = (r - 1)(d + 1)$ and assume that, for every $i \in [r]$ and for every $j \in [n + 1]$, a finite set $X_{i,j} \subset \mathbb{R}^d$ is given (which may be empty). This can be thought of as an r by $(n + 1)$ matrix whose i, j-entry is the set $X_{i,j}$.

$$
\begin{array}{ccccc}
 & G_1 & G_2 & \ldots & G_{n+1} \\
X_1 & X_{1,1} & X_{1,2} & \ldots & X_{1,n+1} \\
X_2 & X_{2,1} & X_{2,2} & \ldots & X_{2,n+1} \\
\vdots & \vdots & \vdots & \vdots & \vdots \\
X_r & X_{r,1} & X_{r,2} & \ldots & X_{r,n+1}
\end{array}
$$

We call the sets $X_i = \bigcup_{j=1}^{n+1} X_{i,j}$ colours and the sets $G_j = \bigcup_{i=1}^{r} X_{i,j}$ groups, (nothing to do with groups in algebra though). A transversal of this system is a set $Y = \{y_1, y_2, \ldots, y_{n+1}\}$ if $y_j \in G_j$ for every j. In the multiset case, of course, every $y_j \in Y$ comes from a uniquely determined $X_{i,j} \subset G_j$. The following result is form Arocha et al. [1].

Theorem 8.2. *Under the above conditions, if every transversal is separated along the colours, then so is some group G_j.*

Proof. We use Sarkaria's lemma. A transversal Y is separated along the colours iff $0 \notin \text{conv}\,\overline{Y}$ (where $\overline{Y} \in \mathbb{R}^n$). If $0 \notin \text{conv}\,\overline{Y}$ for all transversals, then by Theorem 4.1, we can't have $0 \in \text{conv}\,\overline{G}_j$ for all j, meaning that $0 \notin \text{conv}\,\overline{G}_j$ for some $j \in [n+1]$. Then, by Sarkaria's lemma again, G_j is separated along the colours.

Note that using Theorem 5.1 instead of colourful Carathéodory, one gets a little more, namely, two groups whose union is separated along the colours. $\qquad\square$

We give two applications of this result. The first is the colourful Kirchberger Theorem 8.1. The finite X is partitioned as $X_1 \cup \cdots \cup X_r$ and we define $X_{i,j} = X_i$ for all $j = [n+1]$. A transversal Y in this case is sequence of $n+1$ elements of X (possibly with repetitions), and Y is separated along the colours simply means that $\bigcap_1^r \text{conv}(Y \cap X_i) = \emptyset$. If all transversals are separated along the colours, then so is one group by the theorem we just proved. But all groups are the same, which means that $X_1, \ldots, X_r$ are separated along the colours.

The second application is a proof of Tverberg's theorem. We are given a set $X = \{x_1, \ldots, x_{n+1}\}$ in $\mathbb{R}^d$, $n = (r-1)(d+1)$ and we are going to find an r-partition $X_1 \cup \cdots \cup X_r$ of X with $\bigcap_1^r \text{conv}\,X_i \neq \emptyset$. Define $X_{i,j} = \{x_j\}$ for all $i \in [r]$. As each group is a single point repeated r times, no group is separated along the colours. Theorem 8.2 implies then that some transversal, say Y, is not separated along the colours. Note that each $y_j \in Y$ comes from a unique $X_{i,j}$. For a fixed i, let X_i be the set of $y_j \in Y$ that come from $X_{i,j}$. This is a partition of X. The fact that Y is not separated along the colours means exactly that $\bigcap_1^r \text{conv}\,X_i \neq \emptyset$, as required.

Open question 8.3. Give an effective algorithm to find a Tverberg partition of a set $X \subset \mathbb{R}^d$ with $(r-1)(d+1)+1$ elements. Note that a positive answer to Open question 4.5 would solve this problem, via Sarkaria's lemma.

9 Tverberg's theorem with tolerance

A partition of a finite set $X \subset \mathbb{R}^d$ with parts $X_1, \ldots, X_r$ has tolerance t if for every set $T \subset X$ of size t

$$\bigcap_1^r \text{conv}(X_i \setminus T) \neq \emptyset.$$

A partition with tolerance with $t = 0$ is just a Tverberg partition. The question is what size of X, as a function of d, r and t, guarantees the existence of an r-partition with tolerance t. This question is open even in the case $r = 2, t = 1$ (see [27] for more information). Recently Soberón and Strausz [27] have given an upper bound on this number. Their argument uses Sarkaria's lemma, that is why we present it here.

Theorem 9.1. *Suppose $d \geq 1, r \geq 2, t \geq 0$ are integers. Every $X \subset \mathbb{R}^d$ with at least $(r - 1)(d + 1)(t + 1) + 1$ elements has a partition into r parts with tolerance t.*

Note that the slightly weaker bound $(t + 1)[(r - 1)(d + 1) + 1]$ follows from Tverberg's theorem directly.

It will be convenient to say that $S \subset \mathbb{R}^d$ captures the origin if $0 \in \text{conv } S$, and S captures the origin with tolerance t if $0 \in \text{conv}(S \setminus T)$ for every $T \subset S$ with $|T| \leq t$.

We need a definition and a lemma. Given $S \subset S' \subset \mathbb{R}^p$ and a group G, an action of G on S' is said to be compatible with S if the following holds:

- If $A \subset S'$ captures the origin, then so does gA for every $g \in G$,
- Ga captures the origin for every $a \in S$.

Lemma 9.2. *Assume $p \geq 1$ and $t \geq 0$ are integers, $n = p(t + 1) + 1$, $S = \{a_1, \ldots, a_n\} \subset \mathbb{R}^d$, and G is a finite group with $|G| \leq p$. If there is an action of G on a set S' which is compatible with $S \subset S'$, then there are $g_j \in G$ (for all $j \in [n]$) such that the set $\{g_1a_1, \ldots, g_na_n\}$ captures the origin with tolerance t.*

We prove the theorem first.

Proof of Theorem 9.1. Set $n = (r-1)(d+1)(t+1)+1, X = \{x_1, \ldots, x_n\}$, $p = (r - 1)(d + 1)$, and let $v_1, v_2, \ldots, v_r \in \mathbb{R}^{r-1}$ be the vertices of a regular simplex centered at the origin. So $\alpha_1 v_1 + \cdots + \alpha_r v_r = 0$ iff $\alpha_1 = \cdots = \alpha_r$. Then $v_1, \ldots, v_r$ satisfy the conditions of Sarkaria's lemma. Define

$$S' = \{v_i \otimes (x_j, 1) \in \mathbb{R}^p : i \in [r], j \in [n]\} \text{ and}$$
$$S = \{v_r \otimes (x_j, 1) \in \mathbb{R}^p : j \in [n]\}.$$

There is a natural action of $\mathbb{Z}_r$ (the cyclic group of order r) on S', given by $m(v_i \otimes (x_j, 1)) = v_{i+m} \otimes (x_j, 1)$ where $i + m$ is taken mod r.

Next we check the conditions of Lemma 9.2. For each $a \in S', \mathbb{Z}_r a$ captures the origin as $\sum_1^r v_i = 0$. Suppose A is a subset of S' that captures the origin. As the simplex with vertices $v_1, \ldots, v_r$ is regular, the coefficients of the convex combination that give 0 for A work for gA to give 0 again for every $g \in \mathbb{Z}_r$. Note that the condition $d \geq 1$ is needed as it implies $p \geq r = |G|$.

So the lemma applies and gives $m_j \in \mathbb{Z}_k$, $(j \in [n])$ such that, with $m_j(v_r \otimes (x_j, 1)) = v_{m_j} \otimes (x_j, 1)$ the set

$$Y = \{m_1(v_r \otimes (x_1, 1)), m_2(v_r \otimes (x_2, 1)), \ldots, m_n(v_r \otimes (x_n, 1))\}$$
$$= \{v_{m_1} \otimes (x, 1), \ldots, v_{m_n} \otimes (x_n, 1)\}$$

captures the origin with tolerance t.

We are almost done. Define $X_i = \{x_j : m_j = 1\}$ for $i \in [r]$. This is an r-partition of X and with this partition the set Y is exactly the set $\overline{X}$ that appears in Sarkaria's lemma. As Y captures the origin with tolerance t, for every $T \subset X$ of size at most t, $0 \in \text{conv}(\overline{X \setminus T})$. Sarkaria's lemma implies then that $\bigcap_1^r \text{conv}(X_i \setminus T) \neq \emptyset$. So $X_1, \ldots, X_r$ form an r-partition of X with tolerance t. $\qquad\square$

Proof of the lemma. Let $G = \{g_1, \ldots, g_q\}$, $q \leq p$. We use induction on r. The case $t = 0$ is the colourful Carathéodory Theorem 4.1 with $Ga_1, \ldots, Ga_n$ as colour classes. Suppose the lemma is true for $t - 1$ but false for t. Given a vector $(h_1, \ldots, h_n) \in G^n$ define $h \cdot S = \{h_1 a_1, \ldots, h_n a_n\}$. Since the lemma is false for r, for every $h \in G^n$ there is $T \subset h \cdot S$ with t points so that $h \cdot S \setminus T$ is separated from the origin. So $\text{dist}(0, \text{conv}(h \cdot S \setminus T)) > 0$.

For a given $h \in G^n$ let $D(h)$ denote the minimum of all such distances, so $D(h) > 0$. Choose $h^* \in G^*$ so that $D(h^*)$ is minimal among all the $D(h)$. Let T^* be the t-element subset of $h^* \cdot S$ for which $D(h^*) = \text{dist}(0, \text{conv}(h^* \cdot S \setminus T^*))$. Write $\Delta = \text{conv}(h^* \cdot S \setminus T^*)$, so there is $x \in \Delta$ which realizes this distance. Let H be the hyperplane in $\mathbb{R}^p$ that contains x and is orthogonal to x. It follows that x is in the convex hull of a set $V \subset (h^* \cdot S \setminus T^*) \cap H$ with at most p elements. Write $U = h^* \cdot S \setminus V$ and let H^- be the halfspace bounded by H and containing the origin.

It is easy to see that U is compatible with the action of G, and $m = |U| \geq pt + 1$. The induction hypothesis yields a vector $k \in G^{|U|}$ such that $k \cdot U$ captures the origin with tolerance $t - 1$. Observe that for each $b \in U$ there is $g_i \in G$ such that $g_i b \in H^-$. This follows as the set Gb captures the origin for every $b \in U$. Consider the sets $(g_1 k) \cdot U, \ldots, (g_q k) \cdot U$

written as rows in the matrix below.

$$g_1k_1u_1 \quad g_1k_2u_2 \quad \ldots \quad g_1k_mu_m$$
$$g_2k_1u_1 \quad g_2k_2u_2 \quad \ldots \quad g_2k_mu_m$$
$$\vdots \qquad \vdots \qquad \vdots \qquad \vdots$$
$$g_qk_1u_1 \quad g_qk_2u_2 \quad \ldots \quad g_qk_mu_m$$

By the previous observation, every column here contains an element in H^-. There are $m \geq pt + 1$ columns and $q \leq p$ rows. By the pigeonhole principle there is a $g \in G$ such that $(gk) \cdot U$ has at least $t + 1$ elements in H^-.

Next we define a new vector $h \in G^n$ by setting $h_j = gk_j$ if $a_j \in U$ and $h_j = h_j^*$ otherwise. We claim that $D(h) < D(h^*)$. Let $T \subset h \cdot S$ be a set of at most t points such that $0 \notin \mathrm{conv}((h \cdot S) \setminus T)$. Now T cannot contain $t - 1$ or fewer points from $(gh^*) \cdot U$, because then $h \cdot S \setminus T$ would capture the origin. Thus $T \subset (gh^*) \cdot U$ and then there is a point $a \in H^- \cap ((gh^*) \cdot U)$ that is not in T. It follows that $\mathrm{conv}(V \cup \{a\})$ is closer to the origin than $\mathrm{conv}\, V$. Thus indeed $D(h) < D(h^*)$ contradicting the minimality of $D(h^*)$. $\qquad\square$

Open question 9.3. Write $T(d, r, t)$ for the smallest integer such that every set $X \subset \mathbb{R}^d$ with $T(d, r, t)$ points has an r-partition with tolerance t. Theorem 9.1 shows that $T(r, d, t) \leq (r - 1)(d + 1)(t + 1) + 1$. What is the exact value of $T(d, r, t)$? Even the case $t = 1$ is open, the best known lower bound is $\lfloor 5d/3 \rfloor + 3 \leq T(d, r, 1)$, cf [22] and [16].

Some recent results concerning $T(d, r, t)$ for $d \leq 2$ can be found in Mulzer, Stein [19].

ACKNOWLEDGEMENTS. Thanks are due to Pablo Soberón and the anonymous referees for thorough and critical reading and useful advice. My research was partially supported by ERC Advanced Research Grant no 267165, and by Hungarian National Research Grants K 83767 and NK 78439.

References

[1] J. AROCHA, I. BÁRÁNY, J. BRACHO, R. FABILA and L. MONTE-JANO, *Very colorful theorems*, Discrete Comp. Geom. **42** (2009), 142–154.

[2] I. BÁRÁNY, *A generalization of Carathéodory's theorem*, Discrete Math. **40** (1982), 141–152.

[3] I. BÁRÁNY, Z. FÜREDI and L. LOVÁSZ, *On the number of halving planes*, Combinatorica **10** (1990), 175–185.

[4] I. BÁRÁNY and D. G. LARMAN, *A coloured version of Tverberg's theorem*, J. London Math. Soc. **45** (1992), 314–320.

[5] I. BÁRÁNY and SH. ONN, *Colourful linear programming and its relatives*, Math. OR **22** (1997), 550–567.

[6] I. BÁRÁNY and J. MATOUŠEK, *Quadratic lower bound for the number of colourful simplices*, SIAM J. on Discrete Math. **21** (2007), 191–198.

[7] I. BÁRÁNY and R. KARASEV, *Notes about the Carathéodory number*, Discrete Comp. Geom. **48** (2012), 783–792

[8] A. BJÖRNER, L. LOVÁSZ, R.T. ŽIVALJEVIĆ and S.T. VREĆICA, *Chessboard complexes and matching complexes*, J. London Math. Soc. **45** (1944), 25–39.

[9] P. BLAGOJEVIĆ, B. MATSCHKE and G. M. ZIEGLER, *Optimal bounds for the colored Tverberg problem*, October 2009. (arXiv:math.AT/0910.4987).

[10] A. DEZA, S. HUANG, T. STEPHEN and T. TERLAKY, *Colourful simplicial depth*, Discrete Comp. Geom. **35** (2006), 597–615.

[11] A. DEZA, T. STEPHEN and F. XIE, *More colourful simplices*, Discrete Comp. Geom. **45** (2011), 272–278.

[12] A. DEZA, F. MEUNIER and P. SARRABEZOLLES, *A combinatorial approach to colourful simplicial depth*, SIAM J. Discrete Math. **28** (2014), 306–322.

[13] W. FENCHEL, *Über Krümmung und Windung geschlossener Raumkurven*, Math. Ann. **101** (1929), 238–252.

[14] A. HOLMSEN. J. PACH and H. TVERBERG, *Points surrounding the origin*, Combinatorica **28** (2008), 633–634.

[15] P. KIRCHBERGER, *Über Tschebyschefsche Annäherungsmethoden*, Math. Ann. **57** (1903), 509–540.

[16] D. G. LARMAN, *On sets projectively equivalent to the vertices of a convex polytope*, Bull. London Math. Soc. **4** (1972), 6–12

[17] J. MATOUŠEK, *Note on the colored Tverberg theorem*, J. Comb. Theory B **66** (1966), 146–151.

[18] J. W. MILNOR, "Topology from the Differentiable Viewpoint", The University Press of Virginia, Charlottesville, VA, 1965

[19] W. MULZER and Y. STEIN, *Algorithms for tolerated Tverberg partitions* (2013) arXiv:1036.5527.

[20] A. PÓR, "Diploma thesis", Eötvös University, Budapest, 1998.

[21] J. RADON, *Mengen konvexer Körper, die einen gemensamen Punkt erhalten*, Math. Ann. **83** (1921), 113–115.

[22] J. L. RAMÍREZ ALFONSÍN, *Lawrence oriented matroids and a problem of McMullen on projective equivalences of polytopes*, European J. Comb. **22** (2001), 723–731.

[23] J-P. ROUDNEFF, *Partitions of Points into Simplices with k-dimensional Intersection. Part I: The Conic Tverbergís Theorem*, Europ. J. Comb. **22** (2001), 733–743.

[24] P. SARRABEZOLLES, *The colourful simplicail depth conjecture* (2014), arXiv:1402.3412.

[25] K. S. SARKARIA, *Tverberg's theorem via number fields*, Israel J. Math. **79** (1992), 317–320.

[26] P. SOBERÓN, *Equal coefficients and tolerance in coloured Tverberg partitions*, arXiv:1204.1202

[27] P. SOBERÓN and R. STRAUSZ, *A Generalisation of Tverberg's Theorem*, Discrete Comp. Geom. **47** (2012), 455–460.

[28] T. STEPHEN and H. THOMAS, *A quadratic lower bound for colourful simplicial depth*, J. Comb. Optimization **16** (2008), 324–327.

[29] H. TVERBERG, *A generalization of Radon's theorem*, J. London Math. Soc. **41** (1966), 123–128.

[30] H. TVERBERG, *A generalization of Radon's theorem II*, Bull. Austr. Math. Soc. **24** (1981), 321–325.

[31] H. TVERBERG and S.T. VREĆICA, *On generalizations of Radon's theorem and the Ham sandwich theorem*, European J. Comb. **14** (1993), 259–264.

[32] G. M. ZIEGLER, *3N Colored Points in a Plane*, Notices of the AMS. **58** (2011), 550–557.

[33] R. T. ŽIVALJEVIĆ and S.T. VREĆICA, *The colored Tverberg's problem and complexes of injective functions*, J. Comb. Theory A. **61** (1992), 309–318.

[34] M. YU. ZVAGELSKII, *An elementary proof of Tverberg's theorem*, J. Math. Sci. (N.Y.) **161** (2009), 384–387.

Cliques and stable sets
in undirected graphs

Maria Chudnovsky

Abstract. The *cochromatic number* of a graph G is the minimum number of stable sets and cliques of G covering the vertex-set of G. In this paper we survey some resent results and techniques developed in an attempt to answer the question: excluding which induced subgraphs causes a graph to have bounded cochromatic number?

1 Introduction

All graphs in this paper are finite and simple. Let G be a graph. We denote by $V(G)$ the vertex-set of G. A *tournament* is a directed graph, where for every two vertices u, v exactly one of the (ordered) pairs uv and vu is an edge. For tournaments S and T, we say that T is *S-free* if no subtournament of T is isomorphic to S. If $\mathcal{S}$ is a family of tournaments, then T is *$\mathcal{S}$-free* if T is S-free for every $S \in \mathcal{S}$. A tournament is *transitive* if it has no directed cycles (or, equivalently, no cyclic triangles). For a tournament T, we denote by $\alpha(T)$ the maximum number of vertices of a transitive subtournament of T. Finally, the *chromatic number* of T is the smallest number of transitive subtournaments of T whose vertex-sets have union $V(T)$.

We say that a tournament S is a *hero* if there exists $d > 0$ such that every S-free tournament has chromatic number at most d, and S is a celebrity if there exists $0 < c \leq 1$ such that every S-free tournament T has $\alpha(T) \geq c|V(T)|$. Heroes and celebrities are studied in [1]. Somewhat surprisingly, it turns out that a tournament is a hero if and only if it is a celebrity (the "only if" implication is clear, but the "if" is non-trivial). The main result of [1] says that all heroes (and equivalently celebrities) can be constructed starting from single vertices by repeatedly applying

Partially supported by NSF grants DMS-1001091 and IIS-1117631.

two growing operations; and every tournament constructed in that way is a hero.

Similar questions make sense for undirected graphs as well as tournaments, and the goal of this paper is to survey recent progress on this topic.

2 Heroes without direction

Let G be an undirected graph. For a subset X of $V(G)$ we denote by $G|X$ the subgraph of G induced by X. The *complement* G^c of G is the graph with vertex set $V(G)$, such that two vertices are adjacent in G if and only if they are non-adjacent in G^c. A *clique* in G is a set of vertices all pairwise adjacent. A *stable set* in G is a set of vertices all pairwise non-adjacent (thus a stable set in G is a clique in G^c). The largest size of a clique in G is denoted by $\omega(G)$, and the largest size of a stable set by $\alpha(G)$. The *chromatic number* of G is the smallest number of stable sets of G with union $V(G)$. Given a graph H, we say that G is H-*free* if G has no induced subgraph isomorphic to H. If G is not H-free, we say that G contains H. For a family $\mathcal{F}$ of graphs, we say that G is $\mathcal{F}$-free if G is F-free for every $F \in \mathcal{F}$.

As with tournaments, one might ask what graphs H have the property that every H-free graph G has chromatic number bounded by a constant d (where d depends on H, but not on G). However, this question does not have an interesting answer. The complete graph on n vertices is H-free for every H that is not a complete graph, and has chromatic number n. On the other hand, for every $k > 0$ there exist graphs with no clique of size three and with chromatic number at least k (this is a theorem of Erdős [4]), and so only graphs with at most two vertices have the property.

Let us modify the question a little. The *cochromatic number* of a graph G is the minimum number of stable sets and cliques of G with union $V(G)$. We denote the cochromatic number of G by $co\chi(G)$. Let us say that a family $\mathcal{H}$ is *heroic* if there exists a constant $d(\mathcal{H}) > 0$ such that $co\chi(G) \le d(\mathcal{H})$ for every every $\mathcal{H}$-free graph G, and it is *celebrated* if there exists a constant $0 < c(\mathcal{H}) \le 1$ such that every $\mathcal{H}$-free graph G contains either a clique or a stable set of size at least $c(\mathcal{H})|V(G)|$. Clearly, if $\mathcal{H}$ is heroic, then it is celebrated. This turns out to be a better undirected analogue of the concepts discussed in the Introduction. Heroic and celebrated families of undirected graphs were studied in [3].

Let G be a complete multipartite graph with m parts, each of size m. Then G has m^2 vertices, and no clique or stable set of size larger than m; and the same is true for G^c. Thus every celebrated family contains a complete multipartite graph and the complement of one. The *girth* of a graph is the smallest length of a cycle in it. Recall that for every positive integer g there exist graphs with girth at least g and no linear-size stable set (this is a theorem of Erdős [4]). Consequently, every celebrated family must also contain a graph of girth at least g, and, by taking complements, a graph whose complement has girth at least g. Thus, for a finite family of graphs to be celebrated, it must contain a forest and the complement of one. In particular, if a celebrated family only contains one graph H, then $|V(H)| \leq 2$. The following conjecture, proposed in [3], states that these necessary conditions for a finite family of graphs to be celebrated are in fact sufficient for being heroic.

Conjecture 2.1. A finite family of graphs is heroic if and only if it contains a complete multipartite graph, the complement of a complete multipartite graph, a forest, and the complement of a forest.

We remark that this is closely related to a well-known conjecture made independently by Gyárfás [5] and Sumner [10], that can be restated as follows in the language of heroic families:

Conjecture 2.2. For every complete graph K and every forest T, the family $\{K, T\}$ is heroic.

For partial results on Conjecture 2.2 see [6–9].

Since a complete graph is a multipartite graph, the complement of one, and the complement of a forest, we deduce that Conjecture 2.1 implies Conjecture 2.2. The main result of [3] is that Conjecture 2.1 and Conjecture 2.2 are in fact equivalent. A graph G is c-split if $V(G) = X \cup Y$, where

- $\omega(X) \leq c$, and
- $\alpha(Y) \leq c$.

The fact that Conjecture 2.2 implies Conjecture 2.1 is a consequence of the following theorem of [3]:

Theorem 2.3. *Let K and J be graphs, such that both K and J^c are complete multipartite. Then there exists a constant $c(K, J)$ such that every $\{K, J\}$-free graph is $c(K, J)$-split.*

Now let $\mathcal{F}$ be a family of graphs that contains a complete multipartite graph, the complement of a complete multipartite graph, a forest, and the complement of a forest, and let G be an $\mathcal{F}$-free graph. By Theorem 2.3, there exists a constant c such that G is c-split. Now applying Conjecture 2.2 to $G|X$ and $G^c|Y$, the assertion of Conjecture 2.1 follows.

3 Cographs

The goal of this section is to discuss a generalization of Theorem 2.3. A *cograph* is a graph obtained from one-vertex graphs by repeatedly taking disjoint unions and disjoint unions in the complement. In particular, complete graphs, their complements, complete multipartite graphs, and their complements are all cographs. Thus Theorem 2.3 says that excluding a pair of cographs (a complete multipartite graph and the complement of one) from a graph G guarantees that G has a partition into two parts, each of which excludes a cograph that is in some sense simpler (a complete graph or its complement). It turns out that this idea can be generalized to all cographs.

Let us make this more precise. We say that a graph G is *anticonnected* of G^c is connected. A *component* of G is a maximal non-empty connected subgraph of G, and an *anticomponent* of G is a maximal non-empty anticonnected induced subgraph of G.

First we observe that for every cograph G with at least two vertices, exactly one of G, G^c is connected. Next we recursively define a parameter, called the *height* of a cograph, that measures its complexity. The height of a one-vertex cograph is zero. If G is a cograph that is not connected, let m be the maximum height of a component of G; then the height of G is $m + 1$. If G is a cograph that is not anticonnected, let m be the maximum height of an anticomponent of G; then the height of G is $m + 1$.

Let G be a graph. Given a pair of graphs H_1, H_2, we say that G is $\{H_1, H_2\}$-*split* if $V(G) = X_1 \cup X_2$, where the subgraph of G induced by X_i is H_i-free for every $i \in \{1, 2\}$. One of the results of [2] is the following:

Theorem 3.1. *Let $k > 0$ be an integer, and let H and J be cographs, each of height $k + 1$, such that H is anticonnected, and J is connected. Then there exist cographs $\tilde{H}$ and $\tilde{J}$, each of height k, such that $\tilde{H}$ is connected, and $\tilde{J}$ is anticonnected, and every $\{H, J\}$-free graph is $(\tilde{H}, \tilde{J})$-split.*

Clearly Theorem 2.3 follows from Theorem 3.1 by taking $k = 1$, and observing the cographs of height one are complete graphs and their complements.

4 Excluding pairs of graphs

Given an integer $P > 0$, a graph G, and a set of graphs $\mathcal{F}$, we say that G *admits an* $(\mathcal{F}, P)$-*partition* if the vertex set of G can be partitioned into P subsets $X_1, \ldots, X_P$, so that for every $i \in \{1, \ldots, P\}$, either $|X_i| = 1$, or the subgraph of G induced by X_i is F-free for some $F \in \mathcal{F}$ (we remark that the condition that $|X_i| = 1$ is only necessary when all the members of $\mathcal{F}$ are one-vertex graphs).

The proof of Theorem 2.3 in [3] relies on the following fact:

Theorem 4.1. *Let $p > 0$ be an integer. There exists an integer $r > 0$ such that for every graph G, if every induced subgraph of G with at most r vertices is p-split, then G is p-split.*

Here is a weaker statement that would still imply Theorem 2.3 (here K_p is the complete graph on p vertices, and S_p is the complement of K_p):

Theorem 4.2. *Let $p > 0$ be an integer. There exist integers $r, k > 0$ such that for every graph G, if every induced subgraph of G with at most r vertices is p-split, then G admits a $(\{K_p, S_p\}, k)$-partition.*

However, the proof of Theorem 3.1 did not follow the same route, and no result similar to Theorem 4.2 exists in the setting of general cographs, because of the following theorem of [2]:

Theorem 4.3. *Let H, J be graphs, each with at least one edge. Then for every choice of integers r, k there is a graph G such that*

- *for every $S \subseteq V(G)$ with $|S| \leq r$, the graph $G|S$ is $\{H, J\}$-split, and*
- *G has no $(\{H, J\}, k)$-partition.*

Unfortunately, we do not have an easy construction for Theorem 4.3; our proof involves probabilistic arguments.

By taking complements, the conclusion of Theorem 4.3 also holds if each of H and J has a non-edge. Thus Theorem 4.2 is in a sense the strongest theorem of this form possible. In view of this fact, the proof of Theorem 3.1 in [2] takes a different route, and uses a much more general result, which roughly says that excluding a pair of graphs, one of which is not connected and the other not anticonnected, causes a graph to "break apart" into a bounded number of simpler pieces.

Here is a more precise statement. We denote by $c(H)$ the set of components of H, and by $ac(H)$ the set of anticomponents of H. We remark that for every non-null graph G, at least one of $c(G)$ or $ac(G)$ equals $\{G\}$.

Theorem 4.4. *For every pair of graphs (H, J) there exists an integer P such that every $\{H, J\}$-free graph admits a $(c(H) \cup ac(J), P)$-partition.*

Please note that Theorem 4.4 is trivial unless H is not connected and J is not anticonnected. Even though Theorem 4.4 was motivated by trying to prove Theorem 3.1, its generality makes it an interesting result on its own (possibly more so than Theorem 3.1).

To deduce Theorem 3.1 from Theorem 4.4 one just needs to observe the following:

Theorem 4.5. *Let P, k be positive integers. Let $\mathcal{F}$ be a set of connected cographs, all of height at most k. Then there exists a connected cograph C of height k such that for every partition $X_1, \ldots, X_P$ of $V(C)$ there exists $i \in \{1, \ldots, P\}$ such that $C|X_i$ contains every member of $\mathcal{F}$.*

The proof of Theorem 4.5 is not difficult, and can be found in [2].

5 Back to tournaments

Given tournaments H_1 and H_2 with disjoint vertex sets, we write $H_1 \Rightarrow H_2$ to mean the tournament H with $V(H) = V(H_1) \cup V(H_2)$, and such that $H|V(H_i) = H_i$ for $i = 1, 2$, and every vertex of $V(H_1)$ is adjacent to (rather than from) every vertex of $V(H_2)$. One of the results of [1] is a complete characterization of all heroes. An important and the most difficult step toward that is the following:

Theorem 5.1. *If H_1 and H_2 are heroes, then so is $H_1 \Rightarrow H_2$.*

It turns out that translating the proof of Theorem 4.4 into the language of tournaments gives the following result [2]:

Theorem 5.2. *Let H_1, H_2 be non-null tournaments, and let H be $H_1 \Rightarrow H_2$. Let $m = \max(|V(H_1)|, |V(H_2)|)$. Then every H-free tournament admits an $(\{H_1, H_2\}, 2(m + 1)^m)$-partition.*

Theorem 5.2 immediately implies Theorem 5.1. More precisely, we have:

Theorem 5.3. *Let H_1, H_2 be non-null tournaments, and let H be $H_1 \Rightarrow H_2$. Assume that for $i = 1, 2$ every every H_i-free tournament has chromatic number at most d_i. Let $m = \max(|V(H_1)|, |V(H_2)|)$ and let $d = \max(d_1, d_2)$. Then every H-free tournament has chromatic number at most $2(m + 1)^m d$.*

We remark that this proof of Theorem 5.1 is much simpler than the one in [1].

ACKNOWLEDGEMENTS. This paper is based on results obtained in joint work with Alex Scott and Paul Seymour. The author thanks Irena Penev for her careful reading of the manuscript, and many helpful suggestions.

References

[1] E. BERGER, K. CHOROMANSKI, M. CHUDNOVSKY, J. FOX, M. LOEBL, A. SCOTT, P. SEYMOUR and S. THOMASSÉ, *Tournaments and colouring*, Journal of Combinatorial Theory, Ser. B **103** (2013), 1–20.

[2] M. CHUDNOVSKY, A. SCOTT and P. SEYMOUR, *Excluding pairs of graphs*, Journal of Combinatorial Theory, Ser B, to appear.

[3] M. CHUDNOVSKY and P. SEYMOUR, *Extending the Gyárfás-Sumner conjecture*, Journal of Combinatorial Theory, Ser B, to appear.

[4] P. ERDŐS, *Graph theory and probability*, Canad. J. Math. **11** (1959), 34–38.

[5] A. GYÁRFÁS, *On Ramsey covering-numbers*, Coll. Math. Soc. János Bolyai, In: "Infinite and Finite Sets", North Holland/ American Elsevier, New York, 1975, 10.

[6] A. GYÁRFÁS, E. SZEMEREDI and ZS. TUZA, *Induced subtrees in graphs of large chromatic number*, Discrete Mathematics **30** (1980), 235–244.

[7] H. A. KIERSTEAD and S. G. PENRICE, *Radius two trees specify –bounded classes*, Journal of Graph Theory **18** (1994),119–129.

[8] H. A. KIERSTEAD and Y. ZHU, *Radius three trees in graphs with large chromatic number*, SIAM J. Discrete Math. **17** (2004), 571–581.

[9] A. SCOTT, *Induced trees in graphs of large chromatic number*, Journal of Graph Theory **24** (1997), 297–311.

[10] D. P. SUMNER, *Subtrees of a graph and chromatic number*, In: "The Theory and Applications of Graphs", G. Chartrand (ed.), John Wiley & Sons, New York (1981), 557–576.

A taste of nonstandard methods in combinatorics of numbers

Mauro Di Nasso

Abstract. *By presenting the proofs of a few sample results, we introduce the reader to the use of nonstandard analysis in aspects of combinatorics of numbers.*

Introduction

In the last years, several combinatorial results about sets of integers that depend on their asymptotic densities have been proved by using the techniques of nonstandard analysis, starting from the pioneering work by R. Jin (see *e.g.* [6,8,9,12–14,16,17]). Very recently, the hyper-integers of nonstandard analysis have also been used in Ramsey theory to investigate the partition regularity of possibly non-linear diophantine equations (see [6,19]).

The goal of this paper is to give a soft introduction to the use of nonstandard methods in certain areas of density problems and Ramsey theory. To this end, we will focus on a few sample results, aiming to give the flavor of how and why nonstandard techniques could be successfully used in this area.

Grounding on nonstandard definitions of the involved notions, the presented proofs consist of arguments that can be easily followed by the intuition and that can be taken at first as heuristic reasonings. Subsequently, in the last foundational section, we will outline an algebraic construction of the hyper-integers, and give hints to show how those nonstandard arguments are in fact rigorous ones when formulated in the appropriate language. We will also prove that all the nonstandard definitions presented in this paper are actually equivalent to the usual "standard" ones.

Two disclaimers are in order. Firstly, this paper is not to be taken as a comprehensive presentation of nonstandard methods in combinatorics, but only as a taste of that area of research. Secondly, the presented results are only examples of "first-level" applications of the nonstandard machinery; for more advanced results one needs higher-level nonstandard tools, such as saturation and Loeb measure, combined with other non-elementary mathematical arguments.

1   The hyper-numbers of nonstandard analysis

This introductory section contains an informal description of the basics of nonstandard analysis, starting with the hyper-natural numbers. Let us stress that what follows are not rigorous definitions and results, but only informal discussions aimed to help the intuition and provide the essential tools to understand the rest of the paper.[1]

One possible way to describe the hyper-natural numbers $^*\mathbb{N}$ is the following:

- The *hyper-natural numbers* $^*\mathbb{N}$ are the natural numbers when seen with a "telescope" which allows to also see infinite numbers beyond the usual finite ones. The structure of $^*\mathbb{N}$ is essentially the same as $\mathbb{N}$, in the sense that $^*\mathbb{N}$ and $\mathbb{N}$ cannot be distinguished by any "elementary property".

Here by *elementary property* we mean a property that talks about elements but *not* about subsets[2], and where no use of the notion of infinite or finite number is made.

In consequence of the above, the order structure of $^*\mathbb{N}$ is clear. After the usual finite numbers $\mathbb{N} = \{1, 2, 3, \ldots\}$, one finds the infinite numbers $\xi > n$ for all $n \in \mathbb{N}$. Every $\xi \in {}^*\mathbb{N}$ has a successor $\xi + 1$, and every non-zero $\xi \in {}^*\mathbb{N}$ has a predecessor $\xi - 1$.

$$^*\mathbb{N} = \Big\{ \underbrace{1, 2, 3, \ldots, n, \ldots}_{\text{finite numbers}} \quad \underbrace{\ldots, N-2, N-1, N, N+1, N+2, \ldots}_{\text{infinite numbers}} \Big\}$$

Thus the set of finite numbers $\mathbb{N}$ has not a greatest element and the set of infinite numbers $\mathbb{N}_\infty = {}^*\mathbb{N} \setminus \mathbb{N}$ has not a least element, and hence $^*\mathbb{N}$ is *not* well-ordered. Remark that being a well-ordered set is not an "elementary property" because it is about subsets, not elements.[3]

---

[1]   A model for the introduced notions will be constructed in the last section.

[2]   In logic, this kind of properties are called *first-order* properties.

[3]   In logic, this kind of properties are called *second-order* properties.

- The *hyper-integers* $^*\mathbb{Z}$ are the discretely ordered ring whose positive part is the semiring $^*\mathbb{N}$.
- The *hyper-rationals* $^*\mathbb{Q}$ are the ordered field of fractions of $^*\mathbb{Z}$.

Thus $^*\mathbb{Z} = -^*\mathbb{N} \cup \{0\} \cup {^*\mathbb{N}}$, where $-^*\mathbb{N} = \{-\xi \mid \xi \in {^*\mathbb{N}}\}$ are the negative hyper-integers. The hyper-rational numbers $\zeta \in {^*\mathbb{Q}}$ can be represented as ratios $\zeta = \frac{\xi}{\nu}$ where $\xi \in {^*\mathbb{Z}}$ and $\nu \in {^*\mathbb{N}}$.

As the next step, one considers the hyper-real numbers, which are instrumental in nonstandard calculus.

- The *hyper-reals* $^*\mathbb{R}$ are an ordered field that properly extends both $^*\mathbb{Q}$ and $\mathbb{R}$. The structures $\mathbb{R}$ and $^*\mathbb{R}$ satisfy the same "elementary properties".

As a proper extension of $\mathbb{R}$, the field $^*\mathbb{R}$ is *not* Archimedean, *i.e.* it contains non-zero *infinitesimal* and *infinite* numbers. (Recall that a number ε is infinitesimal if $-1/n < \varepsilon < 1/n$ for all $n \in \mathbb{N}$; and a number Ω is infinite if $|\Omega| > n$ for all n.) In consequence, the field $^*\mathbb{R}$ is *not* complete: *e.g.*, the bounded set of infinitesimals has not a least upper bound.[4]

Each set $A \subseteq \mathbb{R}$ has its *hyper-extension* $^*A \subseteq {^*\mathbb{R}}$, where $A \subseteq {^*A}$. E.g., one has the set of hyper-even numbers, the set of hyper-prime numbers, the set of hyper-irrational numbers, and so forth. Similarly, any function $f : A \to B$ has its *hyper-extension* $^*f : {^*A} \to {^*B}$, where $^*f(a) = f(a)$ for all $a \in A$. More generally, in nonstandard analysis one considers hyper-extensions of arbitrary sets and functions.

The general principle that hyper-extensions are indistinguishable from the starting objects as far as their "elementary properties" are concerned, is called *transfer principle*.

- *Transfer principle*: An "elementary property" P holds for the sets $A_1, \ldots, A_k$ and the functions $f_1, \ldots, f_h$ if and only if P holds for the corresponding hyper-extensions:

$$P(A_1, \ldots, A_k, f_1, \ldots, f_h) \iff P(^*A_1, \ldots, {^*A_k}, {^*f_1}, \ldots, {^*f_h})$$

Remark that all basic set properties are elementary, and so $A \subseteq B \Leftrightarrow {^*A} \subseteq {^*B}$, $A \cup B = C \Leftrightarrow {^*A} \cup {^*B} = {^*C}$, $A \setminus B = C \Leftrightarrow {^*A} \setminus {^*B} = {^*C}$, and so forth.

---

[4] Remark that the property of completeness is *not* elementary, because it talks about subsets and not about elements of the given field. Also the Archimedean property is *not* elementary, because it requires the notion of *finite* hyper-natural number to be formulated.

As direct applications of *transfer* one obtains the following facts: The hyper-rationals $^*\mathbb{Q}$ are *dense* in the hyper-reals $^*\mathbb{R}$; every hyper-real number $\xi \in {}^*\mathbb{R}$ has an an *integer part*, *i.e.* there exists a unique hyper-integer $\mu \in {}^*\mathbb{Z}$ such that $\mu \le \xi < \mu + 1$; and so forth.

As our first example of nonstandard reasoning, let us see a proof of König's Lemma, one of the oldest results in infinite combinatorics.

Theorem 1.1 (König's Lemma – 1927). *If a finite branching tree has infinitely many nodes, then it has an infinite branch.*

Nonstandard proof.   Given a finite branching tree T, consider the sequence of its finite levels $\langle T_n \mid n \in \mathbb{N} \rangle$, and let $\langle T_\nu \mid \nu \in {}^*\mathbb{N} \rangle$ be its hyper-extension. By the hypotheses, it follows that all finite levels $T_n \ne \emptyset$ are nonempty. Then, by *transfer*, also all "hyper-levels" T_ν are nonempty. Pick a node $\tau \in T_\nu$ for some infinite ν. Then $\{t \in T \mid t \le \tau\}$ is an infinite branch of T.   □

2   Piecewise syndetic sets

A notion of largeness used in combinatorics of numbers is the following.

- A set of integers A is *thick* if it includes arbitrarily long intervals:

$$\forall n \in \mathbb{N} \, \exists x \in \mathbb{Z} \, [x, x + n) \subseteq A.$$

In the language of nonstandard analysis:

Definition 2.1 (Nonstandard). *A* is *thick* if $I \subseteq {}^*A$ for some infinite interval I.

By *infinite interval* we mean an interval $[\nu, \mu] = \{\xi \in {}^*\mathbb{Z} \mid \nu \le \xi \le \mu\}$ with infinitely many elements or, equivalently, an interval whose length $\mu - \nu + 1$ is an infinite number.

Another important notion is that of syndeticity. It stemmed from dynamics, corresponding to finite return-time in a discrete setting.

- A set of integers A is *syndetic* if it has bounded gaps:

$$\exists k \in \mathbb{N} \, \forall x \in \mathbb{Z} \, [x, x + k) \cap A \ne \emptyset.$$

So, a set is syndetic means that its complement is not thick. In the language of nonstandard analysis:

Definition 2.2 (Nonstandard). *A is* syndetic *if* $^*A \cap I \neq \emptyset$ *for every infinite interval* I.

The fundamental structural property considered in Ramsey theory is that of partition regularity.

- A family $\mathcal{F}$ of sets is *partition regular* if whenever an element $A \in \mathcal{F}$ is finitely partitioned $A = A_1 \cup \ldots \cup A_n$, then at least one piece $A_i \in \mathcal{F}$.

Remark that the family of syndetic sets fails to be partition regular.[5] However, a suitable weaking of syndeticity satisfies the property.

- A set of integers A is *piecewise syndetic* if $A = T \cap S$ where T is thick and S is syndetic; *i.e.*, A has bounded gaps on arbitrarily large intervals:

$$\exists k \in \mathbb{N} \; \forall n \in \mathbb{N} \; \exists y \in \mathbb{Z} \; \forall x \in \mathbb{Z} \; [x, x+k) \subseteq [y, y+n) \Rightarrow$$
$$\Rightarrow [x, x+k) \cap A \neq \emptyset.$$

In the language of nonstandard analysis:

Definition 2.3 (Nonstandard). *A is* piecewise syndetic *(PS for short) if there exists an infinite interval* I *such that* $^*A \cap I$ *has only finite gaps, i.e.* $^*A \cap J \neq \emptyset$ *for every infinite subinterval* $J \subseteq I$.

Several results suggest the notion of piecewise syndeticity as a relevant one in combinatorics of numbers. *E.g.*, the sumset of two sets of natural numbers having positive density is piecewise syndetic[6]; every piecewise syndetic set contains arbitrarily long arithmetic progressions; a set is piecewise syndetic if and only if it belongs to a minimal idempotent ultrafilter[7].

Theorem 2.4. *The family of PS sets is partition regular.*

---

[5] *E.g.*, consider the partition of the integers determined by

$$A = \bigcup_{n \in \mathbb{N}} [-2^{2n}, -2^{2n-1}) \cup \bigcup_{n \in \mathbb{N}} [2^{2n-1}, 2^{2n})$$

and its complement $\mathbb{Z} \setminus A$, neither of which is syndetic.

[6] This is *Jin's theorem*, proved in 2000 by using nonstandard analysis (see [13]).

[7] See [11, Section 4.4].

Nonstandard proof. By induction, it is enough to check the property for 2-partitions. So, let us assume that $A = \text{BLUE} \cup \text{RED}$ is a PS set; we have to show that RED or BLUE is PS. We proceed as follows:

- Take the hyper-extensions $^*A = {}^*\text{BLUE} \cup {}^*\text{RED}$.
- By the hypothesis, we can pick an infinite interval I where $^*A$ has only finite gaps.
- If the *blue elements of $^*A$ have only finite gaps in I, then BLUE is piecewise syndetic.
- Otherwise, there exists an infinite interval $J \subseteq I$ that only contains *red elements of $^*A$. But then $^*\text{RED}$ has only finite gaps in J, and hence RED is piecewise syndetic. $\qquad\square$

3   Banach and Shnirelmann densities

An important area of research in number theory focuses on combinatorial properties of sets which depend on their density. Recall the following notions:

- The *upper asymptotic density* $\overline{d}(A)$ of a set $A \subseteq \mathbb{N}$ is defined by putting:
$$\overline{d}(A) = \limsup_{n \to \infty} \frac{|A \cap [1, n]|}{n}.$$

- The *upper Banach density* $\text{BD}(A)$ of a set of integers $A \subseteq \mathbb{Z}$ generalizes the upper density by considering arbitrary intervals in place of just initial intervals:
$$\text{BD}(A) = \lim_{n \to \infty} \left(\max_{x \in \mathbb{Z}} \frac{|A \cap [x + 1, x + n]|}{n} \right)$$
$$= \inf_{n \in \mathbb{N}} \left\{ \max_{x \in \mathbb{Z}} \frac{|A \cap [x + 1, x + n]|}{n} \right\}.$$

In order to translate the above definitions in the language of nonstandard analysis, we need to introduce new notions.

In addition to hyper-extensions, a larger class of well-behaved subsets of $^*\mathbb{Z}$ that is considered in nonstandard analysis is the class of *internal* sets. All sets that can be "described" without using the notions of finite or infinite number are internal. Typical examples are the intervals

$$[\xi, \zeta] = \{x \in {}^*\mathbb{Z} \mid \xi \le x \le \zeta\}; \quad [\xi, +\widehat{\infty}) = \{x \in {}^*\mathbb{Z} \mid \xi \ge x\}; \quad etc.$$

Also finite subsets $\{\xi_1, \ldots, \xi_n\} \subset {}^*\mathbb{Z}$ are internal, as they can be described by simply giving the (finite) list of their elements. Internal subsets of ${}^*\mathbb{Z}$ share the same "elementary properties" of the subsets of $\mathbb{Z}$. E.g., every nonempty internal subset of ${}^*\mathbb{Z}$ that is bounded below has a least element; in consequence, the set $\mathbb{N}_\infty$ of infinite hyper-natural numbers is *not* internal. Internal sets are closed under unions, intersections, and relative complements. So, also the set of finite numbers $\mathbb{N}$ is *not* internal, as otherwise $\mathbb{N}_\infty = {}^*\mathbb{N} \setminus \mathbb{N}$ would be internal.

Internal sets are either *hyper-infinite* or *hyper-finite*; for instance, all intervals $[\xi, +\infty)$ are hyper-infinite, and all intervals $[\xi, \zeta]$ are hyper-finite. Every nonempty hyper-finite set $A \subset {}^*\mathbb{Z}$ has its *internal cardinality* $\|A\| \in {}^*\mathbb{N}$; for instance $\|[\xi, \zeta]\| = \zeta - \xi + 1$. Internal cardinality and the usual cardinality agree on finite sets.

If $\xi, \zeta \in {}^*\mathbb{R}$ are hyperreal numbers, we write $\xi \sim \zeta$ when ξ and ζ are *infinitely close*, *i.e.* when their distance $|\xi - \zeta|$ is infinitesimal. Remark that if $\xi \in {}^*\mathbb{R}$ is finite (*i.e.*, not infinite), then there exists a unique real number $r \sim \xi$, namely $r = \inf\{x \in \mathbb{R} \mid x > \xi\}$.[8]

We are finally ready to formulate the definitions of density in nonstandard terms.

Definition 3.1 (Nonstandard). For $A \subseteq \mathbb{N}$, its *upper asymptotic density* $\overline{d}(A) = \beta$ is the greatest real number β such that there exists an infinite $\nu \in {}^*\mathbb{N}$ with

$$\|{}^*A \cap [1, \nu]\|/\nu \sim \beta.$$

Definition 3.2 (Nonstandard). For $A \subseteq \mathbb{Z}$, its *upper Banach density* $BD(A) = \beta$ is the greatest real number β such that there exists an infinite interval I with

$$\|{}^*A \cap I\|/\|I\| \sim \beta.$$

Another notion of density that is widely used in number theory is the following.

- The *Schnirelmann density* $\sigma(A)$ of a set $A \subseteq \mathbb{N}$ is defined by

$$\sigma(A) = \inf_{n \in \mathbb{N}} \frac{|A \cap [1, n]|}{n}.$$

---

[8] Such a real number r is usually called the *standard part* of ξ.

Clearly $BD(A) \geq \overline{d}(A) \geq \sigma(A)$, and it is easy to find examples where inequalities are strict. Remark that $\sigma(A) = 1 \Leftrightarrow A = \mathbb{N}$, and that $BD(A) = 1 \Leftrightarrow A$ is thick. Moreover, if A is piecewise syndetic then $BD(A) > 0$, but not conversely.

Let us now recall a natural notion of embeddability for the combinatorial structure of sets:[9]

- We say that X is *finitely embeddable* in Y, and write $X \leq_{\text{fe}} Y$, if every finite $F \subseteq X$ has a shifted copy $t + F \subseteq Y$.

It is readily seen that transitivity holds: $X \leq_{\text{fe}} Y$ and $Y \leq_{\text{fe}} Z$ imply $X \leq_{\text{fe}} Z$. Notice that a set is $\leq_{\text{fe}}$-maximal if and only if it is thick. Finite embeddability preserves fundamental combinatorial notions:

- If $X \leq_{\text{fe}} Y$ and X is PS, then also Y is PS.
- If $X \leq_{\text{fe}} Y$ and X contains an arithmetic progression of length k, then also Y contains an arithmetic progression of length k.
- If $X \leq_{\text{fe}} Y$ then $BD(X) \leq BD(Y)$.

Remark that while piecewise syndeticity is preserved under $\leq_{\text{fe}}$, the property of being syndetic is *not*. Similarly, the upper Banach density is preserved or increased under $\leq_{\text{fe}}$, but upper asymptotic density is *not*.

Other properties that suggest finite embeddability as a useful notion are the following:

- If $X \leq_{\text{fe}} Y$ then $X - X \subseteq Y - Y$;
- If $X \leq_{\text{fe}} Y$ and $X' \leq_{\text{fe}} Y'$ then $X - X' \leq_{\text{fe}} Y - Y'$.

In the nonstandard setting, $X \leq_{\text{fe}} Y$ means that a shifted copy of the whole X is found in the hyper-extension $^*Y$.

Definition 3.3 (Nonstandard). $X \leq_{\text{fe}} Y$ if $\nu + X \subseteq {}^*Y$ for a suitable $\nu \in {}^*\mathbb{N}$.

Remark that the key point here is that the shift ν could be an infinite number.

The sample result that we present below, due to R. Jin [12], allows to extend results that hold for sets with positive Schnirelmann density to sets with positive upper Banach density.

---

[9] This notion is implicit in I.Z. Ruzsa's paper [20], and has been explicitly considered in [6, Section 4]. As natural as it is, it is well possible that finite embeddability has been also considered by other authors, but I am not aware of it.

Theorem 3.4. *Let $BD(A) = \beta > 0$. Then there exists a set $E \subseteq \mathbb{N}$ with $\sigma(E) \geq \beta$ and such that $E \leq_{fe} A$.*

Nonstandard proof. By the nonstandard definition of Banach density, there exists an infinite interval I such that the relative density $\|^*A \cap I\|/\|I\| \sim \beta$. By translating if necessary, we can assume without loss of generality that $I = [1, M]$ where $M \in \mathbb{N}_\infty$. By a straight counting argument, we will prove the following:

- **Claim.** *For every $k \in \mathbb{N}$ there exists $\xi \in [1, M]$ such that for all $i = 1, \ldots, k$, the relative density $\|^*A \cap [\xi, \xi + i)\|/i \geq \beta - 1/k$.*

We then use an important principle of nonstandard analysis, namely:

- *Overflow*: If $A \subseteq {}^*\mathbb{N}$ is internal and contains all natural numbers, then it also contains all hyper-natural numbers up to an infinite ν:

$$A \text{ internal } \& \mathbb{N} \subset A \implies \exists \nu \in \mathbb{N}_\infty \, [1, \nu] \subseteq A.$$

By the Claim, the internal set below includes $\mathbb{N}$:

$$A = \{\nu \in {}^*\mathbb{N} \mid \exists \xi \in [1, M] \, \forall i \leq \nu \, \|^*A \cap [\xi, \xi + i)\|/i \geq \beta - 1/\nu\}.$$

Then, by *overflow*, there exists an infinite $\nu \in {}^*\mathbb{N}$ and $\xi \in [1, M]$ such that $\|^*A \cap [\xi, \xi + i)\|/i \geq \beta - 1/\nu$ for all $i = 1, \ldots, \nu$. In particular, for all finite $n \in \mathbb{N}$, the real number $\|^*A \cap [\xi, \xi + n)\|/n \geq \beta$ because it is not smaller than $\beta - 1/\nu$, which is infinitely close to β. If we denote by $E = \{n \in \mathbb{N} \mid \xi + n \in {}^*A\}$, this means that $\sigma(E) \geq \beta$. The thesis is reached because $\xi + E \subseteq {}^*A$, and hence $E \leq_{fe} A$, as desired.

We are left to prove the Claim. Given k, assume by contradiction that for every $\xi \in [1, M]$ there exists $i \leq k$ such that $\|^*A \cap [\xi, \xi + i)\| < i \cdot (\beta - 1/k)$. By "hyper-induction" on ${}^*\mathbb{N}$, define $\xi_1 = 1$, and $\xi_{s+1} = \xi_s + n_s$ where $n_s \leq k$ is the least natural number such that $\|^*A \cap [\xi_s, \xi_s + n_s)\| < n_s \cdot (\beta - 1/k)$; and stop at step N when $M - k \leq \xi_N < M$. Since k is finite, we have $k/M \sim 0$ and $\xi_N/M \sim 1$. Then:

$$\beta \sim \frac{1}{M} \cdot \|^*A \cap [1, M]\| \sim \frac{1}{M} \cdot \|^*A \cap [\xi_1, \xi_N)\|$$

$$= \frac{1}{M} \cdot \sum_{s=1}^{N-1} \|^*A \cap [\xi_s, \xi_{s+1})\|$$

$$< \frac{1}{M} \cdot \left(\sum_{s=1}^{N-1} n_s \cdot \left(\beta - \frac{1}{k} \right) \right) = \frac{\xi_N - 1}{M} \cdot \left(\beta - \frac{1}{k} \right) \sim \beta - \frac{1}{k},$$

a contradiction. □

The previous theorem can be strengthened in several directions. For instance, one can find E to be "densely" finitely embedded in A, in the sense that for every finite $F \subseteq X$ one has "densely-many" shifted copies included in Y, *i.e.* $\mathrm{BD}\left(\{t \in \mathbb{Z} \mid t + F \subseteq Y\}\right) > 0$.[10]

4 Partition regularity problems

In this section we focus on the use of hyper-natural numbers in partition regularity problems.

The notion of partition regularity for families of sets given in Section 2, is sometimes weakened as follows:

- A family $\mathcal{F}$ of sets is *weakly partition regular* on X if for every finite partition $X = C_1 \cup \ldots \cup C_n$ there exists $F \in \mathcal{F}$ which is contained in one piece $F \subseteq C_i$.

Differently from the usual approach to nonstandard analysis, here it turns out useful to work in a framework where hyper-extensions can be iterated, so that one can consider, *e.g.*:

- The hyper-hyper-natural numbers $^{**}\mathbb{N}$;
- The hyper-extension $^*\xi \in {}^{**}\mathbb{N}$ of an hyper-natural number $\xi \in {}^*\mathbb{N}$;

and so forth. We remark that working with iterated hyper-extensions requires caution, because of the existence of different levels of extensions.[11] Here, it will be enough to notice that, by *transfer*, one has that $^*\mathbb{N} \subsetneq {}^{**}\mathbb{N}$, and if $\xi \in {}^*\mathbb{N} \setminus \mathbb{N}$ then $^*\xi \in {}^{**}\mathbb{N} \setminus {}^*\mathbb{N}$; and similarly for n-th iterated hyper-extensions.[12]

Let us start with a nonstandard proof of the classic Ramsey theorem for pairs.

Theorem 4.1 (Ramsey – 1928). *Given a finite colouring* $[\mathbb{N}]^2 = C_1 \cup \ldots \cup C_r$ *of the pairs of natural numbers, there exists an infinite set* H *whose pairs are monochromatic:* $[H]^2 \subseteq C_i$.[13]

---

[10] See [6,9] for more on this topic.

[11] See [7] for a discussion of the foundations of iterated hyper-extensions.

[12] Notice also that $^*\mathbb{N}$ is an initial segment of $^{**}\mathbb{N}$, *i.e.* $\xi < \nu$ for every $\xi \in {}^*\mathbb{N}$ and for every $\nu \in {}^{**}\mathbb{N} \setminus {}^*\mathbb{N}$ (such a property is not used in this paper).

[13] In other words, the family $\mathcal{F} = \{[H]^2 \mid H \text{ infinite}\}$ is weakly partition regular on $[\mathbb{N}]^2$.

Nonstandard proof. Take hyper-hyper-extensions and get the finite coloring

$$[^{**}\mathbb{N}]^2 = {}^{**}([\mathbb{N}]^2) = {}^{**}C_1 \cup \ldots \cup {}^{**}C_r.$$

Pick an infinite $\xi \in {}^*\mathbb{N}$, let i be such that $\{\xi, {}^*\xi\} \in {}^{**}C_i$, and consider the set $A = \{x \in \mathbb{N} \mid \{x, \xi\} \in {}^*C_i\}$. Then $\xi \in \{x \in {}^*\mathbb{N} \mid \{x, {}^*\xi\} \in {}^{**}C_i\} = {}^*A$. Now inductively define the sequence $\{a_1 < a_2 < \ldots < a_n < \ldots\}$ as follows:

- Pick any $a_1 \in A$, and let $B_1 = \{x \in \mathbb{N} \mid \{a_1, x\} \in C_i\}$. Then $\{a_1, \xi\} \in {}^*C_i$ and $\xi \in {}^*B_1$.
- $\xi \in {}^*A \cap {}^*B_1 \Rightarrow A \cap B_1$ is infinite.[14] Then pick $a_2 \in A \cap B_1$ with $a_2 > a_1$.
- $a_2 \in B_1 \Rightarrow \{a_1, a_2\} \in C_i$.
- $a_2 \in A \Rightarrow \{a_2, \xi\} \in {}^*C_i \Rightarrow \xi \in {}^*\{x \in \mathbb{N} \mid \{a_2, x\} \in {}^*C_1\} = {}^*B_2$.
- $\xi \in {}^*A \cap {}^*B_1 \cap {}^*B_2 \Rightarrow$ we can pick $a_3 \in A \cap B_1 \cap B_2$ with $a_3 > a_2$.
- $a_3 \in B_1 \cap B_2 \Rightarrow \{a_1, a_3\}, \{a_2, a_3\} \in C_i$, and so forth.

Then the infinite set $H = \{a_n \mid n \in \mathbb{N}\}$ is such that $[H]^2 \subseteq C_i$.   □

We now give some hints on how iterated hyper-extensions can be used in partition regularity of equations. Recall that:

- An equation $E(X_1, \ldots, X_n) = 0$ is [injectively] *partition regular* over $\mathbb{N}$ if the set of [distinct] solutions is weakly partition regular on $\mathbb{N}$, *i.e.*, for every finite coloring $\mathbb{N} = C_1 \cup \ldots \cup C_r$ one finds [distinct] monochromatic $a_1, \ldots, a_n \in C_i$ such that $E(a_1, \ldots, a_n) = 0$.

A useful nonstandard notion in this context is the following:

Definition 4.2. We say that two hyper-natural numbers $\xi, \zeta \in {}^*\mathbb{N}$ are *indiscernible*, and write $\xi \simeq \zeta$, if they cannot be distinguished by any hyper-extension, *i.e.* if for every $A \subseteq \mathbb{N}$ one has either $\xi, \zeta \in {}^*A$ or $\xi, \zeta \notin {}^*A$.[15]

---

[14] Here we use the fact that the hyper-extension $^*X$ of a set $X \subseteq \mathbb{N}$ contains infinite numbers if and only if X is infinite.

[15] The name "indiscernible" is borrowed from mathematical logic. Recall that in model theory two elements are named *indiscernible* if they cannot be distinguished by any first-order formula.

Notice that indiscernibility coincides with equality on finite numbers, because if $k \in \mathbb{N}$ is finite and $\xi \neq k$, then trivially $k \in \{k\} = {}^*\{k\}$ and $\xi \notin {}^*\{k\}$. Notice also that if $k > 1$ is any natural number, then $k\xi \not\simeq \xi$. Indeed, if A is the set of those natural numbers n with the property that the largest exponent a such that k^a divides n is even, then $\xi \in {}^*A \Leftrightarrow k\xi \notin {}^*A$. A useful property that one can easily prove is the following: "If $\xi \simeq \zeta$, then for every $f : \mathbb{N} \to \mathbb{N}$ one has ${}^*f(\xi) \simeq {}^*f(\zeta)$."

By using the notion of indiscernibility, one can reformulate in nonstandard terms:

Definition 4.3 (Nonstandard). An equation $E(X_1, \ldots, X_n) = 0$ is [injectively] *partition regular* on $\mathbb{N}$ if there exist [distinct] hyper-natural numbers $\xi_1 \simeq \ldots \simeq \xi_n$ such that $E(\xi_1, \ldots, \xi_n) = 0$.

The following result recently appeared in [5].

Theorem 4.4. *The equation $X + Y = Z^2$ is* not *partition regular on $\mathbb{N}$, except for the trivial solution $X = Y = Z = 2$.*

Nonstandard proof. Assume by contradiction that there exist $\alpha \simeq \beta \simeq \gamma$ in ${}^*\mathbb{N}$ such that $\alpha + \beta = \gamma^2$. Notice that α, β, γ are infinite, as otherwise $\alpha = \beta = \gamma = 2$ would be the trivial solution. By the hypothesis of indiscernibility, α, β, γ belong to the same congruence class modulo 5, say $\alpha \equiv \beta \equiv \gamma \equiv i \mod 5$ with $0 \leq i \leq 4$. Now write the numbers in the forms:

$$\alpha = 5^a \cdot \alpha_1 + i; \quad \beta = 5^b \cdot \beta_1 + i; \quad \gamma = 5^c \cdot \gamma_1 + i$$

where $a, b, c > 0$ and $\alpha_1, \beta_1, \gamma_1$ are not divisible by 5. Pick a function $f : \mathbb{N} \to \mathbb{N}$ such that, for $n \geq 5$, the value $f(n)$ is the unique $k \not\equiv 0$ mod 5 such that $n = 5^h k + i$ for suitable $h > 0$ and $0 \leq i \leq 4$. Observe that $\alpha_1, \beta_1, \gamma_1$ are the images under ${}^*f$ of α, β, γ respectively; so, $\alpha \simeq \beta \simeq \gamma$ implies that $\alpha_1 \simeq \beta_1 \simeq \gamma_1$, and therefore $\alpha_1 \equiv \beta_1 \equiv \gamma_1 \equiv j \neq 0$ mod 5.

The equality $\alpha + \beta = \gamma^2$ implies that either $i = 0$ or $i = 2$. Assume first that $i = 0$. In this case $\gamma^2 = 5^{2c}\gamma_1^2$ where $\gamma_1^2 \equiv j^2 \neq 0 \mod 5$. If $a < b$ then $\alpha + \beta = 5^a(\alpha_1 + 5^{b-a}\beta_1)$ where $\alpha_1 + 5^{b-a}\beta_1 \equiv j \neq 0 \mod 5$. It follows that $2c = a \simeq c$, a contradiction. If $a > b$ the proof is similar. If $a = b$ then $\alpha + \beta = 5^a(\alpha_1 + \beta_1)$ where $\alpha_1 + \beta_1 \equiv 2j \neq 0 \mod 5$, and also in this case we would have $2c = a \simeq c$, a contradiction. If $i = 2$ then $\gamma^2 - 4 = 5^c(5^c\gamma_1^2 + 4\gamma_1)$ where $5^c\gamma_1^2 + 4\gamma_1 \equiv 4j \neq 0 \mod 5$. Now, in case $a < b$, one has that $\alpha + \beta - 4 = 5^a(\alpha_1 + 5^{b-a}\beta_1)$ where $\alpha_1 + 5^{b-a}\beta_1 \equiv j \neq 0 \mod 5$, and so it would follow that $5^c\gamma_1^2 + 4\gamma_1 =$

$\alpha_1 + 5^{b-a}\beta_1$. But then we would have $4j \equiv j$, which is not possible because $j \not\equiv 0$. The case $a > b$ is similar. Finally, if $a = b$ then $\alpha + \beta - 4 = 5^a(\alpha_1 + \beta_1)$ where $\alpha_1 + \beta_1 \equiv 2j \not\equiv 0 \mod 5$, and it would follow that $4j \equiv 2j$, again reaching the contradiction $j \equiv 0$.   □

The notion of indiscernibility naturally extends to the iterated hyper-extensions of the natural numbers. *E.g.*, if $\Omega, \Xi \in {}^{**}\mathbb{N}$ then $\Omega \simeq \Xi$ means that for every $A \subseteq \mathbb{N}$ one has either $\Omega, \Xi \in {}^{**}A$ or $\Omega, \Xi \notin {}^{**}A$. Notice that $\alpha \simeq {}^*\alpha$ for every $\alpha \in {}^*\mathbb{N}$.

In the sequel, a fundamental role will be played by the following special numbers.

Definition 4.5. A hyper-natural number $\xi \in {}^*\mathbb{N}$ is *idempotent* if $\xi \simeq \xi + {}^*\xi$.[16]

Recall van der Waerden Theorem: *"Arbitrarily large monochromatic arithmetic progressions are found in every finite coloring of $\mathbb{N}$"*. Here we prove a weakened version about 3-term arithmetic progressions, by showing the partition regularity of a suitable equation.

Theorem 4.6. *The diophantine equation $X_1 - 2X_2 + X_3 = 0$ is injectively partition regular on $\mathbb{N}$, which means that for every finite coloring of $\mathbb{N}$ there exists a non-constant monochromatic 3-term arithmetic progression.*

Nonstandard proof. Pick an idempotent number $\xi \in {}^*\mathbb{N}$. The following three distinct numbers in ${}^{***}\mathbb{N}$ are a solution of the given equation:

$$\nu = 2\xi + 0 + {}^{**}\xi \, ; \quad \mu = 2\xi + {}^*\xi + {}^{**}\xi \, ; \quad \lambda = 2\xi + 2{}^*\xi + {}^{**}\xi.$$

That $\nu \simeq \mu \simeq \lambda$ are indiscernible is proved by a direct computation. Precisely, notice that ${}^*\xi \simeq \xi + {}^*\xi$ by the idempotency hypothesis, and so, for every $A \subseteq \mathbb{N}$ and for every $n \in \mathbb{N}$, we have that

$$ {}^*\xi \in {}^{**}A - n = {}^{**}(A - n) \iff \xi + {}^*\xi \in {}^{**}(A - n).$$

In consequence, the properties listed below are equivalent to each other:

---

16 The name "idempotent" is justified by its characterization in terms of ultrafilters: *"$\xi \in {}^*\mathbb{N}$ is idempotent if and only if the corresponding ultrafilter $\mathfrak{U}_\xi = \{A \subseteq \mathbb{N} \mid \xi \in {}^*A\}$ is idempotent with respect to the "pseudo-sum" operation:*

$$A \in \mathcal{U} \oplus \mathcal{V} \iff \{n \mid A - n \in \mathcal{V}\} \in \mathcal{U}$$

where $A - n = \{m \mid m + n \in A\}$". The algebraic structure $(\beta\mathbb{N}, \oplus)$ on the space of ultrafilters $\beta\mathbb{N}$ and its related generalizations have been deeply investigated during the last forty years, revealing a powerful tool for applications in Ramsey theory and combinatorial number theory (see the comprehensive monography [11]). In this area of research, idempotent ultrafilters are instrumental.

- $2\xi + {}^*\xi + {}^{**}\xi \in {}^{***}A$
- $2\xi \in ({}^{***}A - {}^{**}\xi - {}^*\xi) \cap {}^*\mathbb{N} = {}^*[({}^{**}A - {}^*\xi - \xi) \cap \mathbb{N}]$
- $2\xi \in {}^*\{n \in \mathbb{N} \mid \xi + {}^*\xi \in {}^{**}(A - n)\}$
- $2\xi \in {}^*\{n \in \mathbb{N} \mid {}^*\xi \in {}^{**}(A - n)\}$
- $2\xi \in {}^*[({}^{**}A - {}^*\xi) \cap \mathbb{N}] = ({}^{***}A - {}^{**}\xi) \cap {}^*\mathbb{N}$
- $2\xi + {}^{**}\xi \in {}^{***}A.$

This shows that $\nu \simeq \mu$. The other relation $\mu \simeq \lambda$ is proved in the same fashion.[17]   □

One can elaborate on the previous nonstandard proof and generalize the technique. Notice that the considered elements μ, ν, λ were linear combinations of iterated hyper-extensions of a fixed idempotent number ξ, and so they can be described by the corresponding finite strings of coefficients in the following way:

- $\nu = 2\xi + 0 + {}^{**}\xi \rightsquigarrow \langle 2, 0, 1 \rangle$
- $\mu = 2\xi + {}^*\xi + {}^{**}\xi \rightsquigarrow \langle 2, 1, 1 \rangle$
- $\lambda = 2\xi + 2{}^*\xi + {}^{**}\xi \rightsquigarrow \langle 2, 2, 1 \rangle$

Indiscernibility of such linear combinations is characterized by means of a suitable equivalence relation $\approx$ on the finite strings, so that, *e.g.*, $\langle 2, 0, 1 \rangle \approx \langle 2, 1, 1 \rangle \approx \langle 2, 2, 1 \rangle$.

Definition 4.7. The equivalence $\approx$ between (finite) strings of integers is the smallest equivalence relation such that:

- The empty string $\approx \langle 0 \rangle$.
- $\langle a \rangle \approx \langle a, a \rangle$ for all $a \in \mathbb{Z}$.
- $\approx$ is coherent with *concatenations*, *i.e.*

$$\sigma \approx \sigma' \text{ and } \tau \approx \tau' \implies \sigma^\frown \tau \approx \sigma'^\frown \tau'.$$

So, $\approx$ is preserved by inserting or removing zeros, by repeating finitely many times a term or, conversely, by shortening a block of consecutive equal terms. The following characterization is proved in [7]:

- Let $\xi \in {}^*\mathbb{N}$ be idempotent. Then the following are equivalent:

1. $a_0\xi + a_1{}^*\xi + \ldots + a_k \cdot {}^{k*}\xi \simeq b_0\xi + b_1{}^*\xi + \ldots + b_h \cdot {}^{h*}\xi$

---

[17] Here we actually proved the following result ([3] Th. 2.10): *"Let $\mathcal{U}$ be any idempotent ultrafilter. Then every set $A \in 2\mathcal{U} \oplus \mathcal{U}$ contains a 3-term arithmetic progression"*.

2. $\langle a_0, a_1, \ldots, a_k \rangle \approx \langle b_0, b_1, \ldots, b_h \rangle$.

Recall Rado theorem: *"The diophantine equation $c_1 X_1 + \ldots + c_n X_n = 0$ ($c_i \neq 0$) is partition regular if and only if $\sum_{i \in F} c_i = 0$ for some nonempty $F \subseteq \{1, \ldots, n\}$"*. By using the above equivalence, one obtains a non-standard proof of a modified version of Rado theorem, with a stronger hypothesis and a stronger thesis.

Theorem 4.8. *Let $c_1 X_1 + \ldots + c_n X_n = 0$ be a diophantine equation with $n \geq 3$. If $c_1 + \ldots + c_n = 0$ then the equation is injectively partition regular on $\mathbb{N}$.*

Nonstandard proof. Fix $\xi \in {}^*\mathbb{N}$ an idempotent element, and for simplicity denote by $\xi_i = {}^{i*}\xi$ the i-th iterated hyper-extension of ξ. For arbitrary $a_1, \ldots, a_{n-1}$, consider the following numbers in ${}^{n*}\mathbb{N}$:

$$
\begin{aligned}
\mu_1 &= a_1\xi + a_2\xi_1 + a_3\xi_2 + \ldots \quad + a_{n-2}\xi_{n-3} + a_{n-1}\xi_{n-2} + a_{n-1}\xi_{n-1} \\
\mu_2 &= a_1\xi + a_2\xi_1 + a_3\xi_2 + \ldots \quad + a_{n-2}\xi_{n-3} + 0 \qquad\qquad + a_{n-1}\xi_{n-1} \\
\mu_3 &= a_1\xi + a_2\xi_1 + a_3\xi_2 + \ldots \quad + 0 \qquad\qquad + a_{n-2}\xi_{n-2} + a_{n-1}\xi_{n-1} \\
&\ \vdots \qquad \vdots \qquad \vdots \qquad \vdots \qquad \vdots \qquad \vdots \qquad\qquad \vdots \qquad\qquad\quad \vdots \\
\mu_{n-2} &= a_1\xi + a_2\xi_1 + 0 \quad + a_3\xi_3 + \ldots \quad + a_{n-2}\xi_{n-2} + a_{n-1}\xi_{n-1} \\
\mu_{n-1} &= a_1\xi + 0 \quad + a_2\xi_2 + a_3\xi_3 + \ldots \quad + a_{n-2}\xi_{n-2} + a_{n-1}\xi_{n-1} \\
\mu_n &= a_1\xi + a_1\xi_1 + a_2\xi_2 + a_3\xi_3 + \ldots \quad + a_{n-2}\xi_{n-2} + a_{n-1}\xi_{n-1}
\end{aligned}
$$

Notice that $\mu_1 \simeq \ldots \simeq \mu_n$ because the corresponding strings of coefficients are all equivalent to $\langle a_1, \ldots, a_{n-1} \rangle$. Moreover, it can be easily checked that the μ_is are distinct. To complete the proof, we need to find suitable coefficients $a_1, \ldots, a_{n-1}$ in such a way that $c_1\mu_1 + \ldots + c_n\mu_n = 0$. It is readily seen that this happens if the following conditions are fulfilled:

$$
\begin{cases}
(c_1 + \ldots + c_n) \cdot a_1 = 0 \\
(c_1 + \ldots + c_{n-2}) \cdot a_2 + c_n \cdot a_1 = 0 \\
(c_1 + \ldots + c_{n-3}) \cdot a_3 + (c_{n-1} + c_n) \cdot a_2 = 0 \\
\qquad \vdots \\
c_1 \cdot a_{n-1} + (c_3 + \ldots + c_n) \cdot a_{n-2} = 0 \\
(c_1 + \ldots + c_n) \cdot a_{n-1} = 0
\end{cases}
$$

Finally, observe that the first and last equations are trivially satisfied because of the hypothesis $c_1 + \ldots + c_n = 0$; and the remaining $n -$

2 equations are satisfied by infinitely many choices of the coefficients $a_1, \ldots, a_{n-1}$, which can be taken in $\mathbb{N}$.[18]   □

More results in this direction, including partition regularity of non-linear diophantine equations, have been recently obtained by L. Luperi Baglini (see [19]).

5 A model of the hyper-integers

In this final section we outline a construction for a model where one can give an interpretation to all nonstandard notions and principles that were considered in this paper.

The most used single construction for models of the hyper-real numbers, and hence of the hyper-natural and hyper-integer numbers, is the *ultrapower*.[19] Here we prefer to use the purely algebraic construction of [2], which is basically equivalent to an ultrapower, but where only the notion of quotient field of a ring modulo a maximal ideal is assumed.

- Consider $\mathrm{Fun}(\mathbb{N}, \mathbb{R})$, the ring of real sequences $\varphi : \mathbb{N} \to \mathbb{R}$ where the sum and product operations are defined pointwise.

- Let $\mathfrak{I}$ be the ideal of the sequences that eventually vanish:

$$\mathfrak{I} = \{\varphi \in \mathrm{Fun}(\mathbb{N}, \mathbb{R}) \mid \exists k \, \forall n \geq k \, \varphi(n) = 0\}.$$

- Pick a maximal ideal $\mathfrak{M}$ extending $\mathfrak{I}$, and define the hyper-real numbers as the quotient field:

$$^*\mathbb{R} = \mathrm{Fun}(\mathbb{N}, \mathbb{R})/\mathfrak{M}.$$

- The *hyper-integers* are the subring of $^*\mathbb{R}$ determined by the sequences that take values in $\mathbb{Z}$:

$$^*\mathbb{Z} = \mathrm{Fun}(\mathbb{N}, \mathbb{Z})/\mathfrak{M} \subset {}^*\mathbb{R}.$$

---

[18] Here we actually proved the following result ([7] Th.1.2): *"Let $c_1 X_1 + \ldots + c_n X_n = 0$ be a diophantine equation with $c_1 + \ldots + c_n = 0$ and $n \geq 3$. Then there exists $a_1, \ldots, a_{n-1} \in \mathbb{N}$ such that for every idempotent ultrafilter $\mathcal{U}$ and for every $A \in a_1 \mathcal{U} \oplus \ldots \oplus a_{n-1}\mathcal{U}$ there exist distinct $x_i \in A$ such that $c_1 x_1 + \ldots + c_n x_n = 0$".*

[19] For a comprehensive exposition of nonstandard analysis grounded on the ultrapower construction, see R. Goldblatt's textbook [10].

- For every subset $A \subset \mathbb{R}$, its hyper-extension is defined by:

$$^*A = \text{Fun}(\mathbb{N}, A)/\mathfrak{M} \subset {}^*\mathbb{R}.$$

  So, e.g., the *hyper-natural numbers* $^*\mathbb{N}$ are the cosets $\varphi + \mathfrak{M}$ of sequences $\varphi : \mathbb{N} \to \mathbb{N}$ of natural numbers; the hyper-prime numbers are the cosets of sequences of prime numbers, and so forth.

- For every function $f : A \to B$ (where $A, B \subseteq \mathbb{R}$), its hyper-extension $^*f : {}^*A \to {}^*B$ is defined by putting for every $\varphi : \mathbb{N} \to A$:

$$^*f(\varphi + \mathfrak{M}) = (f \circ \varphi) + \mathfrak{M}.$$

- For every sequence $\langle A_n \mid n \in \mathbb{N} \rangle$ of nonempty subsets of $\mathbb{R}$, its hyper-extension $\langle A_\nu \mid \nu \in {}^*\mathbb{N} \rangle$ is defined by putting for every $\nu = \varphi + \mathfrak{M} \in {}^*\mathbb{N}$:

$$A_\nu = \{ \psi + \mathfrak{M} \mid \psi(n) \in A_{\varphi(n)} \text{ for all } n \} \subseteq {}^*\mathbb{R}.$$

It can be directly verified that $^*\mathbb{R}$ is an ordered field whose positive elements are $^*\mathbb{R}^+ = \text{Fun}(\mathbb{N}, \mathbb{R}^+)/\mathfrak{M}$. By identifying each $r \in \mathbb{R}$ with the coset $c_r + \mathfrak{M}$ of the corresponding constant sequence, one obtains that $^*\mathbb{R}$ is a proper superfield of $\mathbb{R}$. The subset $^*\mathbb{Z}$ defined as above is a discretely ordered ring having all the desired properties.

Remark that in the above model, one can interpret all notions used in this paper. We itemize below the most relevant ones.

Denote by $\alpha = \iota + \mathfrak{M} \in {}^*\mathbb{N}$ the infinite hyper-natural number corresponding to the identity sequence $\iota : \mathbb{N} \to \mathbb{N}$.

- The nonempty *internal sets* $B \subseteq {}^*\mathbb{R}$ are the sets of the form $B = A_\alpha$ where $\langle A_n \mid n \in \mathbb{N} \rangle$ is a sequence of nonempty sets. When all A_n are finite, $B = A_\alpha$ is called *hyper-finite*; and when all A_n are infinite, $B = A_\alpha$ is called *hyper-infinite*.[20]

- If $B = A_\alpha$ is the hyper-finite set corresponding to the sequence of nonempty finite sets $\langle A_n \mid n \in \mathbb{N} \rangle$, then its *internal cardinality* is defined by setting $\|B\| = \vartheta + \mathfrak{M} \in {}^*\mathbb{N}$ where $\vartheta(n) = |A_n| \in \mathbb{N}$ is the sequence of cardinalities.

- If $\varphi, \psi : \mathbb{N} \to \mathbb{Z}$ and the corresponding hyper-integers $\nu = \varphi + \mathfrak{M}$ and $\mu = \psi + \mathfrak{M}$ are such that $\nu < \mu$, then the (internal) interval

---

[20] It is proved that any internal set $A \subseteq {}^*\mathbb{R}$ is either hyper-finite or hyper-infinite.

$[\nu, \mu] \subseteq {}^*\mathbb{Z}$ is defined as A_α where $\langle A_n \mid n \in \mathbb{N} \rangle$ is any sequence of sets such that $A_n = [\varphi(n), \psi(n)]$ whenever $\varphi(n) < \psi(n)$.[21]

In full generality, one can show that the *transfer* principle holds. To show this in a rigorous manner, one needs first a precise definition of "elementary property", which requires the formalism of first-order logic. Then, by using a procedure known in logic as "induction on the complexity of formulas", one proves that the equivalences $P(A_1, \ldots, A_k, f_1, \ldots, f_h) \Leftrightarrow P({}^*A_1, \ldots, {}^*A_k, {}^*f_1, \ldots, {}^*f_h)$ hold for all elementary properties P, sets A_i, and functions f_j.

Remark that all the nonstandard definitions given in this paper are actually equivalent to the usual "standard" ones. As examples, let us prove some of those equivalences in detail.

Let us start with the definition of a *thick set* $A \subseteq \mathbb{Z}$. Assume first that there exists a sequence of intervals $\langle [a_n, a_n + n] \mid n \in \mathbb{N} \rangle$ which are included in A. If $\langle [a_\nu, a_\nu + \nu] \mid \nu \in {}^*\mathbb{N} \rangle$ is its hyper-extension then, by *transfer*, every $[a_\nu, a_\nu + \nu] \subseteq {}^*A$, and hence ${}^*A$ includes infinite intervals. Conversely, assume that A is not thick and pick $k \in \mathbb{N}$ such that for every $x \in \mathbb{Z}$ the interval $[x, x + k] \not\subseteq A$. Then, by *transfer*, for every $\xi \in {}^*\mathbb{Z}$ the interval $[\xi, \xi + k] \not\subseteq {}^*A$, and hence ${}^*A$ does not contain any infinite interval.

We now focus on the nonstandard definition of *upper Banach density*. Let $BD(A) \geq \beta$. Then for every $k \in \mathbb{N}$, there exists an interval $I_k \subset \mathbb{Z}$ of length $|I_k| \geq k$ and such that $|A \cap I_k|/|I_k| > \beta - 1/k$. By *overflow*, there exists an infinite $\nu \in {}^*\mathbb{N}$ and an interval $I \subset {}^*\mathbb{Z}$ of internal cardinality $\|I\| \geq \nu$ such that the ratio $\|{}^*A \cap I\|/\|I\| \geq \beta - 1/\nu \sim \beta$. Conversely, let I be an infinite interval such that $\|{}^*A \cap I\|/\|I\| \sim \beta$. Then, for every given $k \in \mathbb{N}$, the following property holds: "There exists an interval $I \subset {}^*\mathbb{Z}$ of length $\|I\| \geq k$ and such that $\|{}^*A \cap I\|/\|I\| \geq \beta - 1/k$". By *transfer*, we obtain the existence of an interval $I_k \subset \mathbb{Z}$ of length $|I_k| \geq k$ and such that $|A \cap I_k|/|I_k| \geq \beta - 1/k$. This shows that $BD(A) \geq \beta$, and the proof is complete.

Let us now turn to *finite embeddability*. Assume that $X \leq_{\text{fe}} Y$, and enumerate $X = \{x_n \mid n \in \mathbb{N}\}$. By the hypothesis, $\bigcap_{i=1}^n (Y - x_i) \neq \emptyset$ for every $n \in \mathbb{N}$ and so, by *overflow*, there exists an infinite $\mu \in {}^*\mathbb{N}$ such that the hyper-finite intersection $\bigcap_{i=1}^\mu ({}^*Y - x_i) \neq \emptyset$. If ν is any

---

[21] One can prove that this definition is well-posed. Indeed, if $\varphi + \mathfrak{M} < \psi + \mathfrak{M}$ and $\langle A_n \mid n \in \mathbb{N} \rangle$ and $\langle A'_n \mid n \in \mathbb{N} \rangle$ are two sequences of nonempty sets such that $A_n = A'_n$ whenever $\varphi(n) < \psi(n)$, then $A_\alpha = A'_\alpha$.

hyper-integer in that intersection, then $v + X \subseteq {}^*Y$. Conversely, let us assume that $v + X \subseteq {}^*Y$ for a suitable $v \in {}^*\mathbb{Z}$. Then for every finite $F = \{x_1, \ldots, x_k\} \subset X$ one has the elementary property: "$\exists v \in {}^*\mathbb{Z}$ $(v + x_1 \in {}^*Y$ & ... & $v + x_k \in {}^*Y)$". By *transfer*, it follows that "$\exists t \in \mathbb{Z}$ $(t + x_1 \in Y$ & ... & $t + x_k \in Y)$", *i.e.* $t + F \subseteq Y$.[22]

We finish this paper with a few suggestions for further readings. A rigorous formulation and a detailed proof of the *transfer principle* can be found in Chapter 4 of the textbook [10], where the *ultrapower* model is considered.[23] See also Section 4.4 of [4] for the foundations of nonstandard analysis in its full generality. A nice introduction of nonstandard methods for number theorists, including a number of examples, is given in [15] (see also [12]). Finally, a full development of nonstandard analysis can be found in several monographies of the existing literature; see *e.g.* the classical H. J. Keisler's book [18], or the comprehensive collections of surveys in [1].

References

[1] L.O. ARKERYD, N.J. CUTLAND and C.W. HENSON, eds., "Nonstandard Analysis – Theory and Applications", NATO ASI Series C, Vol. 493, Kluwer A.P., 1997.

[2] V. BENCI and M. DI NASSO, *A ring homomorphism is enough to get nonstandard analysis*, Bull. Belg. Math. Soc. **10** (2003), 481–490.

[3] V. BERGELSON and N. HINDMAN, *Non-metrizable topological dynamics and Ramsey theory*, Trans. Am. Math. Soc. **320** (1990), 293–320.

[4] C.C. CHANG and H.J. KEISLER, "Model Theory" (3rd edition), North-Holland, 1990.

[5] P. CSIKVARI, K. GYARMATI and A. SARKOZY, *Density and Ramsey type results on algebraic equations with restricted solution sets*, Combinatorica **32** (2012), 425–449.

---

[22] For the equivalence of the nonstandard definition of partition regularity of an equation, one needs a richer model than the one presented here. Precisely, one needs the so-called $\mathfrak{c}^+$-*enlargement property*, that can be obtained in models of the form ${}^*\mathbb{R} = \text{Fun}(\mathbb{R}, \mathbb{R})/\mathfrak{M}$ where $\mathfrak{M}$ is a maximal ideals of a special kind (see [2]).

[23] Remark that our algebraic model is basically equivalent to an ultrapower. Indeed, for any maximal ideal $\mathfrak{M}$ of the ring $\text{Fun}(\mathbb{N}, \mathbb{R})$, the family $\mathcal{U} = \{Z(\varphi) \mid \varphi \in \mathfrak{M}\}$ where $Z(f) = \{n \in \mathbb{N} \mid \varphi(n) = 0\}$ is an ultrafilter on $\mathbb{N}$. By identifying each coset $\varphi + \mathfrak{M}$ with the corresponding $\mathcal{U}$-equivalence class $[\varphi]$, one obtains that the quotient field $\text{Fun}(\mathbb{N}, \mathbb{R})/\mathfrak{M}$ and the ultrapower $\mathbb{R}^{\mathbb{N}}/\mathcal{U}$ are essentially the same object.

[6] M. DI NASSO, *Embeddability properties of difference sets*, Integers **14** (2014), A27.

[7] M. DI NASSO, *Iterated hyper-extensions and an idempotent ultra-filter proof of Rado's theorem*, Proc. Amer. Math. Soc., to appear.

[8] M. DI NASSO, I. GOLDBRING, R. JIN, S. LETH, M. LUPINI and K. MAHLBURG, *Progress on a sumset conjecture by Erdös*, Arxiv:1307.0767, 2013.

[9] M. DI NASSO, I. GOLDBRING, R. JIN, S. LETH, M. LUPINI and K. MAHLBURG, *High density piecewise syndeticity of sumsets*, Arxiv:1310.5729, 2013.

[10] R. GOLDBLATT, "Lectures on the Hyperreals – An Introduction to Nonstandard Analysis", Graduate Texts in Mathematics, Vol. 188, Springer, 1998.

[11] N. HINDMAN and D. STRAUSS, "Algebra in the Stone-Čech Compactification" (2nd edition), Walter de Gruyter, 2012.

[12] R. JIN, *Nonstandard methods for upper Banach density problems*, J. Number Theory **91** (2001), 20–38.

[13] R. JIN, *The sumset phenomenon*, Proc. Amer. Math. Soc. **130** (2002), 855–861.

[14] R. JIN, *Freimans inverse problem with small doubling property*, Adv. Math. **216** (2007), 711–752.

[15] R. JIN, *Introduction of nonstandard methods for number theorists*, Proceedings of the CANT (2005) – Conference in Honor of Mel Nathanson, Integers **8** (2) (2008), A7.

[16] R. JIN, *Characterizing the structure of A+B when A+B has small upper Banach density*, J. Number Theory **130** (2010), 1785–1800.

[17] R. JIN, *Plunnecke's theorem for asymptotic densities*, Trans. Amer. Math. Soc. **363** (2011), 5059–5070.

[18] H.J. KEISLER, "Elementary Calculus – An Infinitesimal Approach" (2nd edition), Prindle, Weber & Schmidt, Boston, 1986. (This book is now freely downloadable from the author's homepage: `http://www.math.wisc.edu/~keisler/calc.html`.)

[19] L. LUPERI BAGLINI, *Partition regularity of nonlinear polynomials: a nonstandard approach*, Integers **14** (2014), A30.

[20] I.Z. RUZSA, *On difference sets*, Studia Sci. Math. Hungar. **13** (1978), 319–326.

A coding problem for pairs of subsets

Béla Bollobás, Zoltán Füredi, Ida Kantor, Gyula O. H. Katona
and Imre Leader

Abstract. Let X be an n–element finite set, $0 < k \leq n/2$ an integer. Suppose that $\{A_1, A_2\}$ and $\{B_1, B_2\}$ are pairs of disjoint k-element subsets of X (that is, $|A_1| = |A_2| = |B_1| = |B_2| = k$, $A_1 \cap A_2 = \emptyset$, $B_1 \cap B_2 = \emptyset$). Define the distance of these pairs by $d(\{A_1, A_2\}, \{B_1, B_2\}) = \min\{|A_1 - B_1| + |A_2 - B_2|, |A_1 - B_2| + |A_2 - B_1|\}$. This is the minimum number of elements of $A_1 \cup A_2$ one has to move to obtain the other pair $\{B_1, B_2\}$. Let $C(n, k, d)$ be the maximum size of a family of pairs of disjoint k-subsets, such that the distance of any two pairs is at least d.

Here we establish a conjecture of Brightwell and Katona concerning an asymptotic formula for $C(n, k, d)$ for k, d are fixed and $n \to \infty$. Also, we find the exact value of $C(n, k, d)$ in an infinite number of cases, by using special difference sets of integers. Finally, the questions discussed above are put into a more general context and a number of coding theory type problems are proposed.

1 The transportation distance

Let X be a finite set of n elements. When it is convenient we identify it with the set $[n] := \{1, 2, \ldots, n\}$. The family of the k-sets of an underlying set X is denoted by $\binom{X}{k}$. For $0 < k \leq n/2$ let $\mathcal{Y}$ be the family of unordered disjoint pairs $\{A_1, A_2\}$ of k-element subsets of X (that is, $|A_1| = |A_2| = k$, $A_1 \cap A_2 = \emptyset$). The *transportation distance* or *Enomoto-Katona distance* d on $\mathcal{Y}$ is defined by

$$
\begin{aligned}
&d(\{A_1, A_2\}, \{B_1, B_2\}) \\
&= \min\{|A_1 - B_1| + |A_2 - B_2|, |A_1 - B_2| + |A_2 - B_1|\}.
\end{aligned} \tag{1.1}
$$

Z. Füredi was supported in part by the Hungarian National Science Foundation OTKA 104343, and by the European Research Council Advanced Investigators Grant 267195.
I. Kantor was supported by GAČR grant number P201/12/P288 and partially done while this author visited the Rényi Institute.
G. O. H. Katona was supported by the Hungarian National Foundation OTKA NK104183.

In fact, this is an instance of a more general notion. Whenever (Z, ρ) is a metric space, we can define a metric $\rho^{(s)}$ on $Z^{(s)}$, the set of unordered s-tuples from Z, by

$$\rho^{(s)}(\{x_1, \ldots, x_s\}, \{y_1, \ldots, y_s\}) = \min_{\pi \in S_s} \sum_{i=1}^{s} \rho(x_i, y_{\pi(i)}). \tag{1.2}$$

It is not hard to verify that $\rho^{(s)}$ satisfies the triangle inequality, *i.e.*, it really is a metric. The transportation distance defined above is obtained by taking $s = 2$, Z to be the set of k-elements subsets of X and ρ is half of their symmetric difference.

The minimization problem (1.2) (where ρ can be an arbitrary metric) is one of the fundamental combinatorial optimization problems, a so called *assignment problem*, a special case of a more general *Monge-Kantorovich transportation problem* (see, *e.g.*, the monograph [18]).

The transportation distance between finite sets of the same cardinalities is one of the interesting measurements among many different ways to define how two sets differ from each other. In [1], Ajtai, Komlós and Tusnády considered the assignment problem from a different perspective, and determined with high probability the transportation distance between two sets of points randomly chosen in a unit square.

Since the transportation distance is an important notion, especially from the algorithmic point of view, there are monographs and graduate texts about this topic, see, *e.g.*, [18]. It is also mentioned in the *Encyclopedia of Distances* [5] as the "KMMW metric" (page 245 in Chapter 14) or as the "c-transportation distance". Nevertheless, many combinatorial problems are still unsolved. The packing of sets in spherical spaces with large transportation distance will be discussed in [8].

2 Packings and codes

Given a metric space (Z, ρ) and a distance $h > 0$, the *packing number* $\delta(Z, \geq h)$ is the maximum number of elements in Z with pairwise distance at least h.

A (v, k, t) packing $\mathcal{P} \subseteq \binom{[v]}{k}$ is a family of k-sets with pairwise intersections at most $t-1$ (here $v \geq k \geq t \geq 1$). In other words, every t-subset is covered at most once. Its maximum size is denoted by $P(v, k, t)$. Obviously,

$$P(v, k, t) \leq \binom{v}{t} / \binom{k}{t}. \tag{2.1}$$

If here equality holds then $\mathcal{P}$ is called a Steiner system $S(v, k, t)$, or a *t-design* of parameters v, k, t and $\lambda = 1$ (for more definitions concerning symmetric combinatorial structures esp., difference sets, etc. see, *e.g.*, the monograph by Hall [10]). More generally, for a set K of integers, a family $\mathcal{P}$ on v elements is called a (v, K, t)-design (packing) if every t-subset of $[v]$ is contained in exactly one (at most one) member of $\mathcal{P}$ and $|P| \in K$ for every $P \in \mathcal{P}$.

Determining the packing number is a central problem of Coding Theory, it is essentially the same problem as finding the rate of a large-distance error-correcting code.

If equality holds in (2.1) then every i-subset of $[v]$ is contained in $\binom{v-i}{t-i}/\binom{k-i}{t-i}$ members of $\mathcal{P}$ for $i = 0, 1, \ldots, t - 1$. We say that v, k, and t satisfy the *divisibility conditions* if these t fractions are integers. It was recently proved by Keevash [13] that for any given k and t there exists a bound $v_0(k, t)$ such that these trivial necessary conditions are also sufficient for the existence of a t-design.

$$\text{An } S(v, k, t) \text{ exists if } v, k, \text{ and } t \text{ satisfy} \tag{2.2}$$
$$\text{the divisibility conditions and } v > v_0(k, t).$$

This implies Rödl's theorem [17], that for given k and t as $v \to \infty$

$$P(v, k, t) = (1 + o(1)) \binom{v}{t} / \binom{k}{t}. \tag{2.3}$$

Even more, (2.2) implies that here the error term is only $O(v^{t-1})$. The case $t = 2$ was proved much earlier by Wilson [19]. For this case he also proved the following more general version. For a finite K there exists a bound $v_0(K, 2)$ such that for $v > v_0(K, 2)$

$$\text{a } (v, K, 2) \text{ design exists if } v \text{ and } K \text{ satisfy} \tag{2.4}$$
$$\text{the generalized divisibility conditions,}$$

namely, g.c.d.$(\binom{k}{2} : k \in K)$ divides $\binom{v}{2}$ and g.c.d.$(k - 1 : k \in K)$ divides $v - 1$.

3 Packing pairs of subsets

In this paper, we concentrate on the space $\mathcal{Y}$ of pairs of *disjoint* k-subsets. We say that a set $\mathcal{C} \subset \mathcal{Y}$ of such pairs is a 2-(n, k, d)–*code* if the distance of any two elements is at least d. Let $C(n, k, d)$ be the maximum size of a 2-(n, k, d)-code. Enomoto and Katona in [6] proposed the problem of determining $C(n, k, d)$. For the origin of the problem see [4]. Connections to Hamilton cycles in the Kneser graph $K(n, k)$ are discussed in [12]. The problem makes sense only when $d \leq 2k \leq n$.

It is obvious, that a maximal 2-$(n, k, 1)$ code consists of all the pairs, $C(n, k, 1) = |\mathcal{Y}| = \frac{1}{2}\binom{n}{k}\binom{n-k}{k}$. A 2-$(n, k, 2k)$ code consists of mutually disjoint k-sets, hence $C(n, k, 2k) = \lfloor n/2k \rfloor$.

In Section 5 we present a method for the determination the exact value of $C(n, k, 2k-1)$ for infinitely many n. However, we were able to complete the cases $k = 2, 3$ only, the cases of pairs and triple systems.

Theorem 3.1. *If* $n \equiv 1 \mod 8$ *and* $n > n_0$ *then* $C(n, 2, 3) = \frac{n(n-1)}{8}$. *If* $n \equiv 1, 19 \mod 342$ *and* $n > n_0$ *then* $C(n, 3, 5) = \frac{n(n-1)}{18}$.

The following theorem was proved in [2]. Let $d \le 2k \le n$ be integers. Then

$$C(n, k, d) \le \frac{1}{2} \frac{n(n-1)\cdots(n-2k+d)}{k(k-1)\cdots\lceil\frac{d+1}{2}\rceil \cdot k(k-1)\cdots\lfloor\frac{d+1}{2}\rfloor}. \qquad (3.1)$$

Quisdorff [16] gave a new proof and using ideas from classical coding theory he significantly improved the upper bound for small values of n (for $n \le 4k$). For completeness, in Section 6 we reprove (3.1) in an even more streamlined way.

Concerning larger values of n one can build a 2-(n, k, d) code from smaller ones using the following observation. If $|(A_1 \cup A_2) \cap (B_1 \cup B_2)| \le 2k - d$ holds for the disjoint pairs $\{A_1, A_2\} \in \mathcal{Y}$, $\{B_1, B_2\} \in \mathcal{Y}$ then $d(\{A_1, A_2\}, \{B_1, B_2\}) \ge d$. Take a $(2k-d+1)$-packing $\mathcal{P}$ on n elements and choose a 2-$(|P|, k, d)$-code on each members $P \in \mathcal{P}$. We obtain

$$\sum_{P \in \mathcal{P}} C(|P|, k, d) \le C(n, k, d). \qquad (3.2)$$

This gives

$$P(n, p, 2k-d+1)C(p, k, d) \le C(n, k, d). \qquad (3.3)$$

Fix p (and k, t and d) then Rödl's theorem (2.3) gives

$$(1 + o(1)) \binom{n}{2k-d+1}\binom{p}{2k-d+1}^{-1} C(p, k, d) \le C(n, k, d).$$

Rearranging we get, that the sequence $C(n, k, d)/\binom{n}{2k-d+1}$ is essentially nondecreasing in n, for any fixed p (and k, t and d)

$$C(p, k, d)/\binom{p}{2k-d+1} \le (1 + o(1))C(n, k, d)/\binom{n}{2k-d+1}.$$

Since, obviously, $C(2k,k,d) \geq 1$ we obtain that $\lim\limits_{n \to \infty} C(n,k,d) / (\,_{2k-d+1}^{\quad n}\,)$ exists, it is positive, it equals to its supremum, and finite by (3.1).

It was conjectured ([2], Conjecture 8) that the upper estimate (3.1) is asymptotically sharp. We prove this conjecture in Section 7.

Theorem 3.2.

$$\lim_{n \to \infty} \frac{C(n, k, d)}{n^{2k-d+1}} = \frac{1}{2} \frac{1}{k(k-1) \cdots \lceil \frac{d+1}{2} \rceil \cdot k(k-1) \cdots \lfloor \frac{d+1}{2} \rfloor}.$$

4   The case $d = 2$, the exact values of $C(n, k, 2)$

Besides the cases mentioned in the previous Section (the cases $d = 1$, $d = 2k$ and $(k, d) \in \{(2, 3), (3, 5)\}$) we can solve one more case easily, namely if $d = 2$. Since $C(2k, k, 2) = |\mathcal{Y}| = \frac{1}{2}(\,_{k}^{2k}\,)$ the construction (3.3) gives $P(n, 2k, 2k - 1)\frac{1}{2}(\,_{k}^{2k}\,) \leq C(n, k, 2)$. Then the recent result of Keevash (2.2) gives the lower bound in the following Proposition. The upper bound follows from (3.1).

Proposition 4.1. $C(n, k, 2) = (\,_{2k-1}^{\quad n}\,)\frac{1}{4k}(\,_{k}^{2k}\,)$ for all $n > n_0(k)$ whenever the divisibility conditions of (2.2) hold. $\square$

5   The case $d = 2k - 1$, the exact values of $C(n, k, 2k - 1)$

The distance $\delta(a, b)$ of two integers mod m $(1 \leq a, b \leq m)$ is defined by

$$\delta(a, b) = \min\{|b - a|, |b - a + m|\}.$$

(Imagine that the integers $1, 2, \ldots, m$ are listed around the cirle clockwise uniformly. Then $\delta(a, b)$ is the smaller distance around the circle from a to b.) $\delta(a, b) \leq \frac{m}{2}$ is trivial. Observe that $b - a \equiv d - c$ mod m implies $\delta(a, b) = \delta(c, d)$.

We say that the pair $S = \{s_1, \ldots, s_k\}$, $T = \{t_1, \ldots, t_k\} \subset \{1, \ldots, m\}$ of disjoint sets is *antagonistic* mod m if

(i) all the $k(k - 1)$ integers $\delta(s_i, s_j)$ $(i \neq j)$ and $\delta(t_i, t_j)$ $(i \neq j)$ are different,

(ii) the k^2 integers $\delta(s_i, t_j)$ $(1 \leq i, j \leq k)$ are all different and

(iii) $\delta(s_i, t_j) \neq \frac{m}{2}$ $(1 \leq i, j \leq k)$.

If there is a pair of disjoint antagonistic k-element subsets mod m then $2k^2 + 1 \leq m$ must hold by (ii) and (iii).

Problem 5.1. Is there a pair of disjoint, antagonistic k-element sets mod $2k^2 + 1$?

We have an affirmative answer only in three cases.

Proposition 5.2. *There is a pair of disjoint, antagonistic k-element sets* mod $2k^2 + 1$ *when* $k = 1, 2, 3$.

Proof. We simply give such k-element sets in these cases. It is easy to check that they satisfy the conditions.

$k = 1$: $S = \{1\}$, $T = \{2\}$.
$k = 2$: $S = \{1, 8\}$, $T = \{2, 3\}$.
$k = 3$: $S = \{1, 5, 19\}$, $T = \{2, 13, 15\}$.   □

Lemma 5.3. *If there is a pair of disjoint, antagonistic k-element sets* mod m *then* $C(m, k, 2k - 1) \geq m$.

Proof. Let (S, T) be the antagonistic pair. The shifts $S(u) = \{a + u \bmod m : s \in S\}$, $T(u) = \{s + u \bmod m : s \in T\}(0 \leq u < m)$ will serve as pairs of disjoint subsets of X.

Suppose that $S(u)$ and $S(v)$ $(u \neq v)$ have two elements in common: $s_1 + u = s_2 + v \neq s_3 + u = s_4 + v$ where $s_1, s_2, s_3, s_4 \in S$, $(s_1, s_2) \neq (s_3, s_4)$. The difference is $s_1 - s_2 = s_3 - s_4$ contradicting (i). One can prove in the same way that $T(u)$ and $T(v)$ $(u \neq v)$ and $S(u)$ and $T(v)$, respectively, have at most one element in common. In other words the intersection of any pair from the sets $S(u)$, $T(u)$, $S(v)$, $T(v)$ has at most one element.

Suppose now that both $S(u) \cap S(v)$ and $T(u) \cap T(v)$ are non-empty for some $u \neq v$. Then $s_1 + u = s_2 + v, t_1 + u = t_2 + v$ holds for some $s_1, s_2 \in S, t_1, t_2 \in T$. This leads to $v - u = s_1 - s_2 = t_1 - t_2$, contradicting (i), again.

Finally, suppose that both $S(u) \cap T(v)$ and $T(u) \cap S(v)$ are non-empty for some $u \neq v$. Then $s_1 + u = t_1 + v, t_2 + u = s_2 + v$ is true for some $s_1, s_2 \in S, t_1, t_2 \in T$. Here $v - u = s_1 - t_1 = t_2 - s_2$ is obtained, contradicting either (ii) or (iii) (the latter one, if $s_1 - t_1 = t_1 - s_1$ is obtained).

This proves that the distance of the pairs $(S(u), T(u))$ and $(S(v), T(v))$ $(u \neq v)$ is at least $2k - 1$.   □

Corollary 5.4. *Suppose that there is Steiner family* $S(n, 2k^2 + 1, 2)$ *and a disjoint, antagonistic pair of k-element subsets* mod $2k^2 + 1$ *then*

$$C(n, k, 2k - 1) = \frac{n(n - 1)}{2k^2}.$$

Proof. The upper bound $C(n, k, 2k - 1) \leq n(n - 1)/2k^2$ is a corollary of (3.1).

The lower estimate is obtained from (3.3). By Lemma 5.3 one can choose $2k^2 + 1$ pairs of disjoint k-subsets with distance $2k - 1$ in a set of $2k^2 + 1$ elements. This can be done in each of the members of $\mathcal{S}(n, 2k^2 + 1, 2)$. Since the members have at most one common element, the distance of two pairs in distinct members of $\mathcal{S}(n, 2k^2 + 1, 2)$ will have distance at least $2k - 1$. Therefore all the

$$|\mathcal{S}(n, 2k^2 + 1, 2)|(2k^2 + 1) = \frac{\binom{n}{2}}{\binom{2k^2+1}{2}}(2k^2 + 1) = \frac{n(n - 1)}{2k^2}$$

pairs have distance at least 1.   □

Proof of Theorem 3.1. We only need lower bounds, *i.e.*, constructions. The case $k = 3$ follows from Wilson's theorem (2.2) of the existence of $S(n, 19, 2)$, Proposition 5.2 and Corollary 5.4.

Similarly, the case $k = 2$ for $n \equiv 1, 9 \bmod 72$ follows in the same way using Steiner systems $S(n, 9, 2)$ and the fact $C(9, 2, 3) = 9$ from Corollary 5.4. However, one can see that $C(17, 3, 2) = 34$ and then the results follows from Wilson's theorem (2.4) of the existence of $S(n, \{9, 17\}, 2)$ for all large $n \equiv 1 \bmod 8$ and construction (3.2).

The construction for $C(17, 2, 3)$ is similar to the proof of Lemma 5.3. The 9 pairs there are defined as $\{\{x + 1, x + 8\}, \{x + 2, x + 3\}\} : x \in Z_9\}$. These correspond to a perfect edge decomposition of K_9 into C_4's with side lengths $1, 3, 4$, and 2. For $n = 17$ we take the pairs $\{\{x + 7\}, \{x + 2, x + 6\}\} : x \in Z_{17}\}$ and $\{\{y, y + 11\}, \{y + 7, y + 8\}\} : y \in Z_{17}\}$ which correspond to C_4's of side lengths $(2, 5, 1, 6)$ and $(7, 4, 3, 8)$, respectively.   □

Note that the method gives that $C(n, 1, 1) = \frac{n(n-1)}{2}$ when $n \equiv 1, 3 \bmod 6$. This, however, is trivial for all n.

6   A new proof of the upper estimate

The upper estimate in (3.1) was proved in [2]. We give a new, more illuminating proof here.

Given a pair $\{A, B\}$ of disjoint k-element sets let $\mathcal{P}(\{A, B\}, u, v)$ denote the family of pairs $\{U, V\}$ where $|U| = u$, $|V| = v$ and $U \subseteq A$, $V \subseteq B$ or viceversa. We have

$$|\mathcal{P}(\{A, B\}, u, v)| = 2\binom{k}{u}\binom{k}{v}.$$

Suppose first $u < v$. Then the total number of pairs $\{U, V\}$, $U \cap V = \emptyset$, $|U| = u$, $|V| = v$ in an n-element set is

$$\binom{n}{u}\binom{n - u}{v}.$$

Let $\{A_1, B_1\}, \{A_2, B_2\}$ be two pairs with distance at least d, and $u < v$ be two nonnegative integers such that $u + v = 2k - d + 1$. By definition (1.1), $\mathcal{P}(\{A_1, B_1\}, u, v)$ and $\mathcal{P}(\{A_2, B_2\}, u, v)$ are disjoint. We have

$$
C(n, k, d) \leq \frac{\binom{n}{u}\binom{n-u}{v}}{2\binom{k}{u}\binom{k}{v}}
$$

$$
= \frac{n(n-1)\ldots(n-2k+d)}{2k(k-1)\ldots(k-u+1)k(k-1)\ldots(k-v+1)}
$$

(6.1)

for every pair u, v that satisfies the above requirements. If $u = v$, then equality (6.1) holds by similar arguments.

The numerator does not depend on u, and the denominator is maximized when u and v are as close as possible, $i.e.$, for $u = 2k - \lceil \frac{d-1}{2} \rceil$ and $v = 2k - \lfloor \frac{d-1}{2} \rfloor$. Substituting these values, we obtain the upper estimate in (3.1). □

7 Nearly perfect selection

Let $\mathcal{W}$ be the family of pairs $\{U, V\}$ such that $U, V \subseteq [n]$, $U \cap V = \emptyset$, and $|U| + |V| = 2k - d + 1$ holds.
Note that $|\mathcal{W}| = \frac{1}{2} \sum_{0 \leq u \leq 2k-d+1} \binom{n}{u}\binom{n-u}{(2k-d+1)-u}$. For a pair $\{A, B\}$ of disjoint k-element sets, let $\mathcal{P}(\{A, B\})$ denote the family of pairs $\{U, V\} \in \mathcal{W}$ for which $U \subseteq A$ and $V \subseteq B$, or viceversa.

Lemma 7.1. $d(\{A_1, B_1\}, \{A_2, B_2\}) \leq d - 1$ holds if and only if $\mathcal{P}(\{A_1, B_1\}) \cap \mathcal{P}(\{A_2, B_2\}) \neq \emptyset$.

Proof. Suppose that $\{U, V\} \in \mathcal{P}(\{A_1, B_1\}) \cap \mathcal{P}(\{A_2, B_2\})$, say $U \subset A_1 \cap A_2$ and $V \subset B_1 \cap B_2$. Then $|A_1 - A_2| \leq k - |U|, |B_1 - B_2| \leq k - |V|$ imply $|A_1 - A_2| + |B_1 - B_2| \leq 2k - |U| - |V| = d - 1$ proving the statement. The other case is analogous.

Conversely, if the distance is at most $d - 1$ then either $|A_1 - A_2| + |B_1 - B_2| \leq d - 1$ or $|A_1 - B_2| + |B_1 - A_2| \leq d - 1$ must hold. Suppose that the first one is true. Then $|A_1 \cap A_2| + |B_1 \cap B_2| \geq 2k - d + 1$ follows. Take $U = A_1 \cap A_2$ and a $V \subseteq B_1 \cap B_2$ such that $|V| = 2k - d + 1 - |U|$. Then $\mathcal{P}(\{A_1, B_1\}) \cap \mathcal{P}(\{A_2, B_2\}) \neq \emptyset$ holds, as claimed. □

We can view the sets $\mathcal{P}(\{A, B\})$ as the edges of a hypergraph on the vertex set $\mathcal{W}$. Let us call this hypergraph $\mathcal{H}$. Then a 2-(n, k, d)-code corresponds to a *matching* in $\mathcal{H}$.

In his celebrated paper [17], Rödl established (2.3) in the following way. He viewed the t-element sets as vertices of a $\binom{k}{t}$-uniform hypergraph $\mathcal{H}_n$ whose edges correspond to the k-element subsets of $[n]$. Equality (2.3) is in fact a statement about the existence of an almost perfect matching in $\mathcal{H}_n$. Using the same key proof idea, a powerful generalization by Frankl and Rödl [7] guarantees the existence of almost perfect matchings in hypergraphs satisfying certain more general conditions. Various generalizations and stronger versions versions were later proved, e.g., by Pippenger and Spencer [15].

A function $t : E(\mathcal{H}) \to \mathbb{R}$ is a *fractional matching* of the hypergraph $\mathcal{H}$ if $\sum_{e \in E(\mathcal{H}); x \in e} t(e) \leq 1$ holds for every vertex $x \in V(\mathcal{H})$. The *fractional matching number*, denoted $\nu^*(\mathcal{H})$ is the maximum of $\sum_{e \in E(\mathcal{H})} t(e)$ over all fractional matchings. If $\nu(\mathcal{H})$ denotes the maximum size of a matching in $\mathcal{H}$, then clearly

$$\nu(\mathcal{H}) \leq \nu^*(\mathcal{H}).$$

Kahn [11] proved that under certain conditions, asymptotic equality holds. Both the hypotheses and the conclusion are in the spirit of the Frankl–Rödl theorem.

Given a hypergraph $\mathcal{H}$ with vertex set $[n]$, a fractional matching t and a subset $W \subseteq [n]$, define $\bar{t}(W) = \sum_{W \subseteq e \in E(\mathcal{H})} t(e)$ and $\alpha(t) = \max\{\bar{t}(\{x, y\}) : x, y \in V(\mathcal{H}), x \neq y\}$. In other words, $\alpha(t)$ is a fractional generalization of the codegree. Let $t(\mathcal{H})$ denote $\sum_{e \in E(\mathcal{H})} t(e)$. We say that $\mathcal{H}$ is *s-bounded* if each of its edges has size at most s.

Theorem 7.2 ([11]). *For every s and every $\varepsilon > 0$ there is a δ such that whenever $\mathcal{H}$ is an s-bounded hypergraph and t a fractional matching with $\alpha(t) < \delta$, then*
$$\nu(\mathcal{H}) > (1 - \varepsilon)t(\mathcal{H}).$$

Proof of Theorem 3.2. In the light of Lemma 7.1 it suffices to verify the conditions of Theorem 7.2 and to produce a fractional matching t of the hypergraph $\mathcal{H}$ of the desired size.

Define a constant weight function $t : E(\mathcal{H}) \to \mathbb{R}$ by

$$t(e) = \frac{\lceil \frac{d-1}{2} \rceil! \lfloor \frac{d-1}{2} \rfloor!}{n^{d-1}}.$$

For a vertex $x = \{U, V\} \in \mathcal{W}$ with $|U| = u$ and $|V| = v$ we have

$$\deg(\{U, V\}) = \binom{n - u - v}{k - u}\binom{n - k - v}{k - v}$$
$$\leq \frac{n^{d-1}}{(k - u)!(k - v)!} \leq \frac{n^{d-1}}{\lceil \frac{d-1}{2} \rceil! \lfloor \frac{d-1}{2} \rfloor!}$$

hence t is indeed a fractional matching. Note that $t(\mathcal{H})$ is is asymptotically equal to the quantity in the statement of the Theorem 3.2.

The hypergraph $\mathcal{H}$ is not regular but s-bounded with $s = \frac{1}{2}\sum_u \binom{k}{u} \cdot \binom{k}{(2k-d+1)-u}$. Here s does not depend on n. For $x, y \in V(\mathcal{H}) = W$ let $\deg(x, y)$ denote the codegree of $x = \{U, V\}$ and $y = \{U', V'\}$, i.e., the number of hyperedges $\mathcal{P}(\{A, B\})$ that contain both x and y. If $U \cup V = U' \cup V'$ (they partition the same $(2k - d + 1)$-element set) then the codegre $\deg(x, y) = 0$. Otherwise, $|U \cup U' \cup V \cup V'| \geq 2k - d + 2$ and $(U \cup U' \cup V \cup V') \subset (A \cup B)$ imply that

$$\deg(\{U, V\}, \{U', V\}) = O(n^{d-2}).$$

Hence $\alpha(t) = \deg(\{U, V\}, \{U', V\}) \cdot t(e) = o(1)$ and Kahn's theorem completes the proof. □

8 s-tuples of sets, q-ary codes

Let $\mathcal{Y}^{(s)}$ be the family of s-tuples of pairwise disjoint k-element subsets of $[n]$. A natural definition of a metric on $\mathcal{Y}^{(s)}$ was already mentioned in the introduction, in equation (1.2). With ρ being half the symmetric difference, the distance is defined as

$$\rho^{(s)}(\{A_1, \ldots, A_s\}, \{B_1, \ldots, B_s\}) = \min_{\pi \in S_s} \sum_{i=1}^{s} |A_i \setminus B_{\pi(i)}|.$$

Let $C_s(n, k, d)$ denote the maximum size of a subfamily S of $\mathcal{Y}^{(s)}$ such that any two elements in S have distance at least d. The proofs presented in Sections 7 and 6 can be easily adapted to determining $C_s(n, k, d)$, as well. The proof of the lower and the upper bounds in Theorem 8.1 is completely analogous to the proofs of inequality (3.1) and Theorem 3.2.

Theorem 8.1.

$$\lim_{n \to \infty} \frac{C_s(n, k, d)}{n^{sk-d+1}} = \frac{1}{s!} \frac{\lceil \frac{d-1}{s} \rceil! \lceil \frac{d-2}{s} \rceil! \cdots \lceil \frac{d-s}{s} \rceil!}{(k!)^s}.$$

Let $\mathcal{Y}_q$ be the set of q-ary vectors of length n and weight k (weight is the number of nonzero entries). Let $A_q(n, d, k)$ be the maximum size of a subset $\mathcal{C} \subseteq \mathcal{Y}_q$ such that $\rho'(u, v) \geq d$ whenever $u, v \in \mathcal{C}$. Here ρ' is the Hamming distance.

With a slightly more technical proof along the same lines, the following can be proven.

Theorem 8.2. *Fix $q \geq 2$, k and d. If d is odd, then, as $n \to \infty$,*

$$A_q(n, d, k) \sim \frac{n^{k-\frac{d-1}{2}} (q-1)^{k-\frac{d-1}{2}} \left(\frac{d-1}{2}\right)!}{k!}.$$

If $d \geq 2$ is even, then, as $n \to \infty$,

$$A_q(n, d, k) \sim \frac{n^{k-\frac{d}{2}+1}(q-1)^{k-\frac{d}{2}+1}\left(\frac{d}{2}-1\right)!}{k!}.$$

To use random methods constructing codes is not a new idea. The best known general bounds for the *covering radius* problems are obtained in this way, see, *e.g.*, [9,14].

We can also consider pairs (or more generally s-tuples) of q-ary vectors of weight k. For simplicity, we will only state the results for pairs here. Define the set $\mathcal{Y}_q^{(2)}$ of pairs $\{u, v\}$ of vectors such that

- $u, v \in \{0, 1, \ldots, q-1\}^n$
- each of u and v has exactly k nonzero entries
- the supports of u and v are disjoint (*i.e.* $u_i = 0$ for all i such that $v_i \neq 0$, and $v_i = 0$ for all i such that $u_i \neq 0$).

Define the distance between these pairs by

$$\delta(\{u, v\}, \{w, z\}) = \min\{\rho'(u, w) + \rho'(v, z), \rho'(u, z) + \rho'(v, w)\}$$

where ρ' is again the Hamming distance.

In the following, $A_q^2(n, d, k)$ will denote the maximum size of a subset $\mathcal{C} \subseteq \mathcal{Y}_q^{(2)}$ such that $\delta(\{u, v\}, \{w, z\}) \geq d$ for any pair $\{u, v\}, \{w, z\}$ of members of $\mathcal{C}$.

Theorem 8.3. *Fix q, d and k. If d is odd and $q \geq 3$, then, as $n \to \infty$,*

$$A_q^2(n, d, k) \sim \frac{1}{2} \cdot \frac{n^{2k-\frac{d-1}{2}} \cdot (q-1)^{2k-\frac{d-1}{2}} \cdot \lfloor\frac{d-1}{4}\rfloor! \lceil\frac{d-1}{4}\rceil!}{(k!)^2}.$$

If $d \geq 2$ is even and $q \geq 2$, then, as $n \to \infty$,

$$A_q^2(n, d, k) \sim \frac{1}{2} \cdot \frac{n^{2k-\frac{d}{2}+1} \cdot (q-1)^{2k-\frac{d}{2}} \cdot \lfloor\frac{d}{4}\rfloor! \left(\lceil\frac{d}{4}\rceil-1\right)!}{(k!)^2}.$$

The distance δ used here is twice the distance defined in Section 1, hence the apparent inconsistency of this result for $q = 2$ with Theorem 3.2.

For $q = 2$ and d odd we have $A_q(n, d, k) = A_q(n, d+1, k)$.

9 Open problems

We believe that for an arbitrary pair of k and d, there are infinitely many n's with equality in inequality (3.1).

10   Further developments

Let us note that since announcing the first version of the present paper Theorem 3.1 has been greatly extended by Chee, Kiah, Zhang and Zhang [3]. They determined the exact value of $C(n, 2, d)$ completely, and for any fixed k the exact value of $C(n, k, 2k - 1)$ for all $n > n_0(k)$ satisfying either $n = 0 \mod k$ or $n = 1 \mod k$ and $n(n - 1) = 0 \mod 2k^2$. Their proofs are different: they use more design theory. However, our Section 5 is still interesting for its own sake and Problem 5.1 is still open.

ACKNOWLEDGEMENTS. The authors are very grateful for the helpful remarks of the referees.

References

[1] M. AJTAI, J. KOMLÓS and G. TUSNÁDY, *On optimal matchings*, Combinatorica **4** (1984), 259–264.

[2] G. BRIGHTWELL and G. O. H. KATONA, *A new type of coding problem*, Studia Sci. Math. Hungar. **38** (2001), 139–147.

[3] YEOW MENG CHEE, HAN MAO KIAH, HUI ZHANG, AND XIANDE ZHANG, *Optimal codes in the Enomoto-Katona space*, Combinatorics, Probability and Computing, to appear. (Preliminary version in Proc. IEEE Intl. Symp. Inform. Theory. IEEE, 2013.)

[4] J. DEMETROVICS, G. O. H. KATONA and A. SALI, *Design type problems motivated by database theory*, J. Statist. Plann. Inference **72** (1998), 149–164. R. C. Bose Memorial Conference (Fort Collins, CO, 1995).

[5] M. M. DEZA and E. DEZA, "Encyclopedia of Distances", Springer, 2nd ed. 2013.

[6] H. ENOMOTO and G. O. H. KATONA, *Pairs of disjoint q-element subsets far from each other*, Electron. J. Combin. **8** (2001), Research Paper 7, 7 pp. (electronic). In honor of Aviezri Fraenkel on the occasion of his 70th birthday.

[7] P. FRANKL and V. RÖDL, *Near perfect coverings in graphs and hypergraphs*, European J. Combin. **6** (1985), 317–326.

[8] Z. FÜREDI, *Packings of sets in spherical spaces with large transportation distance*, in preparation.

[9] Z. FÜREDI and J-H. KANG, Covering the n-space by convex bodies and its chromatic number, Discrete Mathematics **308** (2008), 4495–4500.

[10] M. HALL, "Combinatorial Theory", 2nd ed., Wiley-Interscience, 1998.

[11] J. KAHN, *A linear programming perspective on the Frankl-Rödl-Pippenger theorem*, Random Structures Algorithms **8** (1996), 149–157.

[12] G. O. H. KATONA, *Constructions via Hamiltonian theorems*, Discrete Math. **303** (2005), 87–103.

[13] P. KEEVASH, *The existence of designs*, arxiv.org 1401.3665.

[14] M. KRIVELEVICH, B. SUDAKOV and VAN H. VU, *Covering codes with improved density*, IEEE Trans. Inform. Theory **49** (2003), no. 7, 1812–1815.

[15] N. PIPPENGER and J. SPENCER, *Asymptotic behavior of the chromatic index for hypergraphs*, J. Combin. Theory Ser. A **51** (1989), 24–42.

[16] JÖRN QUISTORFF, *New upper bounds on Enomoto–Katona's coding type problem*, Studia Sci. Math. Hungar. **42** (2005), 61–72.

[17] V. RÖDL, *On a packing and covering problem*, European J. Combin. **6** (1985), 69–78.

[18] C. VILLANI, "Topics in Optimal Transportation", Vol. 58 of Graduate Studies in Mathematics, American Mathematical Society, Providence, RI, 2003.

[19] R. M. WILSON, *An existence theory for pairwise balanced designs. II. The structure of PBD-closed sets and the existence conjectures*, J. Combinatorial Theory Ser. A **13** (1972), 246–273.

String graphs and separators

Jiří Matoušek

Abstract. String graphs, that is, intersection graphs of curves in the plane, have been studied since the 1960s. We provide an expository presentation of several results, including very recent ones: some string graphs require an exponential number of crossings in every string representation; exponential number is always sufficient; string graphs have small separators; and the current best bound on the crossing number of a graph in terms of pair-crossing number. For the existence of small separators, the proof includes generally useful results on approximate flow-cut dualities.

This expository paper was prepared as a material for two courses co-taught by the author in 2013, at Charles University and at ETH Zurich. It aims at a complete and streamlined presentation of several results concerning *string graphs*. This important and challenging class of intersection graphs has traditionally been studied at the Department of Applied Mathematics of the Charles University, especially by Jan Kratochvíl and his students and collaborators.

A major part of the paper is devoted to a separator theorem by Fox and Pach, recently improved by the author, as well as an application of it by Tóth in a challenging problem from graph drawing, namely, bounding the crossing number by a function of the pair-crossing number. This is an excellent example of a mathematical proof with a simple idea but relying on a number of other results from different areas. The proof is presented in full, assuming very little as a foundation, so that the reader can see everything that is involved. A key step is an approximate flow-cut duality from combinatorial optimization and approximation algorithms, whose proof relies on linear programming duality and a theorem on metric embeddings.

Supported by the ERC Advanced Grant No. 267165 and by the project CE-ITI (GACR P202/12/G061).

1   Intersection graphs

The classes IG($\mathcal{M}$). Let $\mathcal{M}$ be a system of sets; we will typically consider systems of geometrically defined subsets of $\mathbb{R}^2$, such as all segments in the plane. We define IG($\mathcal{M}$), the class of **intersection graphs** of $\mathcal{M}$, by

$$\mathrm{IG}(\mathcal{M}) = \Big\{ (V, E) : V \subseteq \mathcal{M}, E = \{\{M, M'\} \in \tbinom{V}{2} : M \cap M' \neq \emptyset\} \Big\}.$$

In words, the vertices of each graph in IG($\mathcal{M}$) are sets in $\mathcal{M}$, and two vertices are connected by an edge if they have a nonempty intersection.

Usually we consider intersection graphs of $\mathcal{M}$ up to isomorphism; *i.e.*, we regard a graph G as an intersection graph of $\mathcal{M}$ if it is merely isomorphic to a graph $G' \in \mathrm{IG}(\mathcal{M})$. In that case we call $V(G') \subseteq \mathcal{M}$ an **$\mathcal{M}$-representation** of G, or just a **representation** of G if $\mathcal{M}$ is understood.

Important examples

- For $\mathcal{M}$ consisting of all (closed) intervals on the real line, we obtain the class of **interval graphs**. This is one of the most useful graph classes in applications. Interval graphs have several characterizations, they can be recognized in linear time, and there is even a detective story *Who Killed the Duke of Densmore?* by Claude Berge (in French; see [4] for English translation) in which the solution depends on properties of interval graphs.

- **Disk graphs**, *i.e.*, intersection graphs of disks in the plane, and **unit disk graphs** have been studied extensively. Of course, one can also investigate intersection graphs of *balls in $\mathbb{R}^d$* for a given d, or of *unit balls*.

- Another interesting class is CONV, the intersection graphs of convex sets in the plane.

- Here we will devote most of the time to the class STRING of **string graphs**, the intersection graphs of simple curves in the plane.

- Another important class is SEG, the **segment graphs**, which are the intersection graphs of line segments in $\mathbb{R}^2$.

Other interesting classes of graphs are obtained by placing various restrictions on the mutual position or intersection pattern of the sets representing the vertices.

For example:

- For an integer $k \geq 1$, k-STRING is the subclass of string graphs consisting of all graphs representable by curves such that every two of them have at most k points of intersection.[1]

- For $k \geq 1$, the class k-DIR consists of the segment graphs possessing a representation in which the segments involved have at most k distinct directions. (So 1-DIR are just interval graphs.)

- The **kissing graphs of circles**, sometimes also called **contact graphs of circles** or **coin graphs**, are disk graphs that admit a representation by disks with disjoint interiors; that is, every two disks either are disjoint or just touch. The beautiful and surprisingly useful **Koebe–Andreev–Thurston theorem** asserts that a graph is a kissing graph of circles if and only if it is planar. While "only if" is easy to see, the "if" direction is highly nontrivial. Here we have just mentioned this gem of a result but we will not discuss it any further.

Typical questions   For each class C of intersection graphs, and in particular, for all the classes mentioned above, one can ask a number of basic questions. Here are some examples:

- How hard, computationally, is the **recognition problem for C**? That is, given an (abstract) graph G, is it isomorphic to a graph in C? For some classes, such as the interval graphs, polynomial-time or even linear-time algorithms have been found, while for many other classes the recognition problem has been shown NP-hard, and sometimes it is suspected to be even harder (not to belong to NP).

- How complicated a representation may be required for the graphs in C? In more detail, we first need to define some reasonable notion of **size** of a representation of a graph in C. Then we ask, given an integer n, what is the maximum, over all n-vertex graphs $G \in C$, of the smallest possible size of a representation of G?
  For example, it is not too difficult to show that each segment graph has a representation in which all of the segments have endpoints with integral coordinates. For such a representation, the size can be defined as the total number of bits needed for encoding all the coordinates of the endpoints.

---

[1] Some authors moreover require that each of the intersection points is a *crossing*, i.e., a point where, locally, one of the edges passes from one side of the second edge to the other (as opposed to a *touching point*).

- Can the chromatic number be bounded in terms of the clique number? It is well known that there are graphs G with clique number $\omega(G) = 2$, *i.e.*, triangle-free, and with chromatic number $\chi(G)$ arbitrarily large. On the other hand, many important classes, such as interval graphs, consist of *perfect graphs*, which satisfy $\omega(G) = \chi(G)$. Some classes $\mathcal{C}$ display an intermediate behavior, namely, $\chi(G) \leq f(\omega(G))$ for all $G \in \mathcal{C}$ and for some function $f \, \mathbb{N} \to \mathbb{N}$; establishing a bound of this kind is often of considerable interest, and then one may ask for the smallest possible f. If $\chi(G)$ cannot be bounded in terms of $\omega(G)$ alone, one may still investigate bounds for $\chi(G)$ in terms of $\omega(G)$ and the number of vertices of G.

- One may also consider two classes of interest, $\mathcal{C}$ and $\mathcal{C}'$, and ask for inclusion relations among them (*e.g.*, whether $\mathcal{C} \subseteq \mathcal{C}'$, or $\mathcal{C}' \subseteq \mathcal{C}$, or even $\mathcal{C} = \mathcal{C}'$). Some relations are quite easy, such as SEG $\subseteq$ CONV $\subseteq$ STRING, but others may be very challenging.

  For example, Scheinermann conjectured in his PhD. thesis in 1984 that all planar graphs are in SEG. It took over 20 years until Chalopin, Gonçalves, and Ochem [8] managed to prove the weaker result that all planar graphs are in 1-STRING, and in 2009 Chalopin and Gonçalves [7] finally established Scheinermann's conjecture.

Thus, it should already be apparent from our short lists of classes and questions that the study of intersection graphs is an area in which it is very easy to produce problems (and exercises). However, instead of trying to survey the area, we will focus on a small number of selected results. Some of them also serve us as a stage on which we are going to show various interesting tools in action.

Exercise 1.1. Prove carefully an assertion made above: every SEG-graph has a representation with all segment endpoints integral. (Hint: check the definition of SEG again and note what it does *not* assume.)

Exercise 1.2. Show that graphs in 100-STRING can be recognized in NP.

2 Basics of string graphs

We begin with a trivial but important observation: all of the complete graphs K_n are string graphs. Hence, unlike classes such as planar graphs, string graphs can be dense and they have no forbidden minors. Moreover, they are not closed under taking minors; thus, the wonderful Robertson–Seymour theory is not applicable.

Another simple observation asserts that every planar graph is a string graph, even 2-STRING. The following picture indicates the proof:

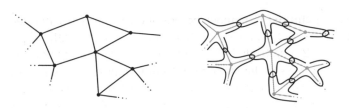

As we have mentioned, every planar graph is even a segment graph, but this is a difficult recent result [7].

Example 2.1. It is not completely easy to come up with an example of a non-string graph. Here is one:

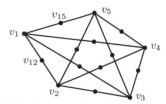

(more generally, every graph obtained from a non-planar graph by replacing each edge by a path of length at least two is non-string).

Sketch of proof. For contradiction we suppose that this graph has a representation by simple curves (referred to as *strings* in this context), where each v_i is represented by a string γ_i and v_{ij} is represented by γ_{ij}. From such a representation we will obtain a planar drawing of K_5, thus reaching a contradiction.

To this end, we first select, for each $i < j$, a piece π_{ij} of γ_{ij} connecting a point of γ_i to a point of γ_j and otherwise disjoint from γ_i and γ_j. Next, we continuously shrink each γ_i to a point, pulling the π_{ij} along — the result is the promised planar drawing of K_5. The picture shows this construction in the vicinity of the string γ_1:

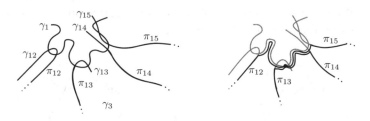

$\square$

Admittedly, this argument is not very rigorous, and if the strings are arbitrary curves, it is difficult to specify the construction precisely. An easier route towards a rigorous proof hinges on the following generally useful fact.

Lemma 2.2. *Every (finite) string graph G can be represented by polygonal curves, i.e., simple curves consisting of finitely many segments. We may also assume that every two curves have finitely many intersection points, and that no point belongs to three or more curves.*

Sketch of proof. We start from an arbitrary string representation of G. By compactness, there exists an $\varepsilon > 0$ such that every two disjoint strings in the representation have distance at least ε. For every two strings γ, δ that intersect, we pick a point $p_{\gamma\delta}$ in the intersection. Then we replace each string γ by a polygonal curve that interconnects all the points $p_{\gamma\delta}$ and lies in the open $\frac{\varepsilon}{2}$-neighborhood of γ.

By a small perturbation of the resulting polygonal curves we can then achieve finitely many intersections and eliminate all triple points.   □

Exercise 2.3. Let $U \subseteq \mathbb{R}^2$ be an open, arcwise connected set; that is, every two points of U can be connected by a simple curve in U. Prove, as rigorously as possible, that every two points of U can also be connected by a polygonal curve.

Let us call a string representation as in Lemma 2.2 **standard**.

3   String graphs requiring exponentially many intersections

How hard is to recognize string graphs?   Using an ingenious reduction, Kratochvíl [19] proved that recognizing string graphs is NP-hard, but the question remained, does this problem belong to the class NP?

A natural way of showing membership of the problem in NP would be to guess a string representation, and verify in polynomial time that it indeed represents a given graph G. A simple way of specifying a string representation is to put a vertex into every intersection point of the strings, and describe the resulting plane (multi)graph:

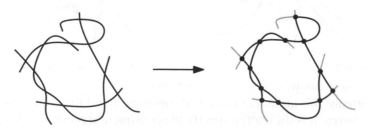

In this description, the edges are labeled by the strings they come from. Then it can be checked whether such a plane graph indeed provides a string representation of G.

This argument may seem to prove membership in NP easily, but there is a catch: namely, we would need to know that there is a polynomial $p(n)$ such that every string graph on n vertices admits a string representation with at most $p(n)$ intersection points. However, as was noticed in [17], this is false—as we will prove below, there are string graphs for which every representation has *exponentially many intersections*. After this result, for ten years it was not clear whether there is *any algorithm at all* for recognizing string graphs.

Weak realizations   As an auxiliary device, we introduce the following notions. An **abstract topological graph** is a pair (G, R), where G is an (abstract) graph and $R \subseteq \binom{E(G)}{2}$ is a symmetric relation on the edge set. A **weak realization** of such (G, R) is a drawing of G in the plane such that whenever two edges e, e' intersect (sharing a vertex does not count), we have $\{e, e'\} \in R$. Thus, R specifies which pairs of edges are allowed (but not forced) to intersect.

We call a weak realization **standard** if the corresponding drawing of G is **standard**, by which we mean that the edges are drawn as polygonal curves, every two intersect at finitely many points, and no three edges have a common intersection (where sharing a vertex does not count). (Moreover, as in every graph drawing we assume that the edges do not pass through vertices.) Standard drawings help us to get rid of "local" difficulties in proofs.

Exercise 3.1. (a) Prove that if (G, R) has a weak realization, then it also has a standard weak realization. (This is analogous to Lemma 2.2, but extra care is needed near the vertices!)

(b) Prove that if (G, R) has a weak realization W with finitely many edge intersections in which no three edges have a common intersection, then it also has a standard weak realization W' with at most as many edge intersections as in W.

For a string graph G, let $f_s(G)$ denote the minimum number of intersection points in a standard string representation of G, and let

$$f_s(n) := \max\{f_s(G) : G \text{ a string graph on } n \text{ vertices}\}.$$

Similarly, for an abstract topological graph (G, R) admitting a weak realization, let $f_w(G, R)$ be the minimum number of edge intersections in a standard weak realization of (G, R), and

$$f_w(m) := \max\{f_w(G, R) : (G, R) \text{ weakly realizable}, |E(G)| = m\}.$$

Observation 3.2. $f_w(m) \leq f_s(2m)$.

Proof. Let (G, R) be an abstract topological graph with m edges witnessing $f_w(m)$. We may assume that G is connected and non-planar (why?), and thus $m \geq n = |V(G)|$.

We consider a (standard) weak realization W of (G, R) with $f_w(m)$ intersections, and construct a string representation of a string graph H as follows: we replace every vertex in W by a tiny *vertex string*, and every edge by an *edge string*, as is indicated below:

This H has $m + n \leq 2m$ vertices, and using the monotonicity of f_s, it suffices to show that $f_s(H) \geq f_w(m)$. This follows since a string representation of H with x intersections yields a weak realization of (G, R) with at most x intersections, by contracting the vertex strings to points and pulling the edge strings along (this is the same argument as in Example 2.1). □

Exercise 3.3. Prove that $f_s(n) \leq f_w(n^2) + n^2$. (Or $f_s(n) \leq f_w(O(n^2)) + O(n^2)$ if this looks easier.)

Theorem 3.4. *There is a constant $c > 0$ such that $f_w(m) \geq 2^{cm}$, and consequently, $f_s(n) \geq 2^{(c/2)n}$.*

Proof. For $k \geq 1$, we define a planar graph P_k according to the following picture:

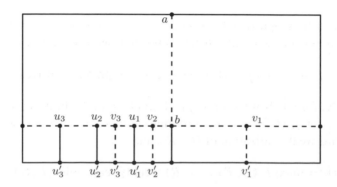

(P_k is obtained from P_{k-1} by adding vertices u_k and v_k to the left and right of u_{k-1}, respectively, and adding the vertical edges $\{u_k, u'_k\}$ and $\{v_k, v'_k\}$). Then we create an abstract topological graph (G_k, R_k) from P_k: G_k is obtained from P_k by adding the edges $\{u_1, v_1\}, \ldots, \{u_k, v_k\}$, and the relation R_k allows each of the edges $\{u_i, v_i\}$ to intersect all of the edges drawn dashed in the picture above. No other edge intersections are permitted.

Each (G_k, R_k) has a weak realization:

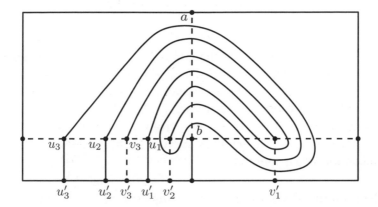

We prove by induction on i that in every weak realization of G_k, the edge $\{u_i, v_i\}$ intersects $\{a, b\}$ at least 2^{i-1} times, $1 \le i \le k$; then the theorem will follow.

Since P_k is a 3-connected graph, it has a topologically unique drawing. From this the case $i = 1$ can be considered obvious. For $i \ge 2$, the situation for the edge $\{u_i, v_i\}$ looks, after contracting the edge $\{u_{i-1}, u'_{i-1}\}$ and a simplification preserving the topology, as follows:

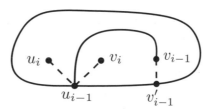

Thus, the edge $\{u_i, v_i\}$ has to cross $\{v_{i-1}, v'_{i-1}\}$.

Now we will use the drawing of $\{u_i, v_i\}$ to get two different curves π_1, π_2 that both "duplicate" the previous edge $\{u_{i-1}, v_{i-1}\}$. The first curve π_1 starts at u_{i-1} and follows $\{u_{i-1}, u_i\}$ up to the point where $\{u_i, v_i\}$ intersects $\{u_{i-1}, u_i\}$ the last time before hitting $\{v_{i-1}, v'_{i-1}\}$ (that point can also be u_i). Then π_1 follows $\{u_i, v_i\}$ almost up to the first intersection with $\{v_{i-1}, v'_{i-1}\}$, and finally, it goes very near $\{v_{i-1}, v'_{i-1}\}$ until v_{i-1}:

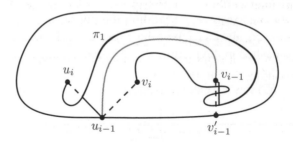

If we remove the drawings of the edges $\{u_j, v_j\}$, $j \geq i - 1$, from the considered weak realization of G_k, and add π_1 as a new way of drawing the edge $\{u_{i-1}, v_{i-1}\}$, we obtain a weak realization of G_{i-1}. Therefore, by the inductive hypothesis, π_1 crosses $\{a, b\}$ at least 2^{i-2} times.

Similarly we construct π_2, disjoint from π_1, which starts at u_{i-1} and first follows $\{u_{i-1}, v_i\}$. It again has to cross $\{a, b\}$ at least 2^{i-2} times, and the induction step is finished. $\square$

4   Exponentially many intersections suffice

The first algorithm for recognizing string graphs was provided by Schaefer and Štefankovič [28], who proved an upper bound on the number of intersections sufficient for a representation of every n-vertex string graph. Similar to the previous section, their proof works with weak representations.

Theorem 4.1 ([28]). *We have $f_w(m) \leq m2^m$. Consequently (by Exercise 3.3), $f_s(n) = 2^{O(n^2)}$.*

This result implies, by the argument given at the beginning of Section 3, that string graphs can be recognized in NEXP (nondeterministic exponential time).

Later Schaefer, Sedgwick, and Štefankovič [29] proved that string graphs can even be recognized in NP. The main idea of their ingenious argument is that, even though a string representation may require exponentially many intersections, there is always a representation admitting a compact encoding, of only polynomial size, by something like a context-free grammar. They also need to show that, given such a compact encoding of a collection of strings, one can verify in polynomial time whether it represents a given graph. We will not discuss their proof any further and we proceed with a proof of Theorem 4.1.

Let (G, R) be a weakly realizable abstract topological graph with m edges. It has a standard weak realization (edges are polygonal curves with finitely many intersections, and no triple intersections; see Exercise 3.1). Moreover, we can make sure that the edges cross at every intersection point, since a "touching point" ⌣╱ can be perturbed away: ⌣╱ (note the advantage of working with weak realizations, in which we need not worry about losing intersections).

Theorem 4.1 is an immediate consequence of the following claim: *if W is a standard weak realization in which some edge e has at least 2^m crossings, then there is another standard weak realization W' with fewer crossing than in W*.

Lemma 4.2. *If an edge e has at least 2^m crossings, then there is a contiguous segment $\hat{e}$ of e that contains at least one crossing and that crosses every edge of G an* even *number of times.*

Proof. The lemma is an immediate consequence of the following combinatorial statement: *If w is a word (finite sequence) of length 2^m over an m-letter alphabet Σ, then there is a nonempty subword (contiguous subsequence) x of w in which each symbol of Σ occurs an even number of times.*

To prove this statement, let us define, for $i = 0, 1, \ldots, 2^m$, a mapping $f_i \Sigma \rightarrow \{0, 1\}$, where $f_i(a) = 0$ if a occurs an even number of times among the first i symbols of w, and $f_i(a) = 1$ otherwise. Since there are only 2^m distinct mappings $\Sigma \rightarrow \{0, 1\}$, there are two indices $i \neq j$ with $f_i = f_j$. Then the subword of w beginning at position $i + 1$ and ending at position j is the desired x. $\square$

Now we fix e and $\hat{e}$ as in the lemma. We can deform the plane by a suitable homeomorphism so that $\hat{e}$ is a horizontal straight segment and there is a narrow band along it, which we call the *window*, in which the edges crossing $\hat{e}$ appear as little vertical segments, and in which no other portions of the edges are present:

Any edge $f \neq e$ has an even number $2n_f$ crossings with $\hat{e}$, and it intersects the border of the window in $4n_f$ points. Let us number these $4n_f$ points as $p_{f,1}, \ldots, p_{f,4n_f}$ in the order as they appear along f (we choose one of the two possible directions of traversing f arbitrarily).

Here is the procedure for redrawing the original weak representation W into W' so that the total number of crossings is reduced.

1. We apply a suitable homeomorphism of the plane that maps the window to a circular disk with $\hat{e}$ as the horizontal diameter, while the edges crossing $\hat{e}$ still appear as vertical segments within the window. For every f and every $i = 1, 2, \ldots, 2n_f - 1$, the point $p_{f,2i}$ is connected to $p_{f,2i+1}$ by an arc of f outside the window. Let us call these arcs for i odd the *odd connectors* and for i even the *even connectors*. The following illustration shows only one edge f, although the window may be intersected by many edges. The points $p_{f,i}$ are labeled only by their indices, and the odd connectors are drawn thick:

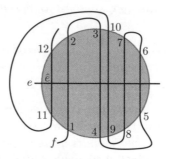

2. We erase everything inside the window. Then we map the odd connectors inside the window by the circular inversion that maps the outside of the window to its inside, while the even connectors stay outside. Crucially, two odd connectors that did not intersect before the circular inversion still do not intersect. Next, we apply the mirror reflection about $\hat{e}$ inside the window to the odd connectors: As the picture il-

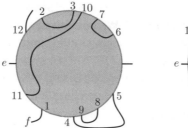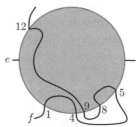

lustrates, these transformed odd connectors together with the original even connectors connect up the initial piece of f to the final piece. This new way of drawing of f crosses the window n_f times, only half of the original number. This is the moment where we use the fact that each edge crosses $\hat{e}$ an even number of times; otherwise, the re-connection of f would not work.

After this redrawing of all the edges crossing $\hat{e}$, edges that did not cross before still do not cross (while some intersections may be lost).

3. It remains to draw the erased portion of e. We do not want to draw it horizontally, since we would have no control over the intersections with the transformed odd connectors. Instead, we draw it along the top or bottom half-circle bounding the window, whichever gives a smaller number of intersections (breaking a tie arbitrarily). Each f crosses

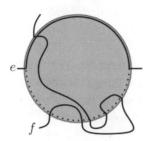

the window border $2n_f$ times after the redrawing, and thus one of the half-circles is crossed at most $\sum_f n_f$ times—while originally $\hat{e}$ was crossed $2\sum_f n_f$ times. Hence the redrawing indeed reduces the number of crossings. The resulting weak realization is not necessarily standard, but we can make it standard without increasing the number of intersections (Exercise 3.1(b)), and Theorem 4.1 is proved.

5   A separator theorem for string graphs

Let G be a graph. A subset $S \subseteq V(G)$ is called a **separator** if there is a partition of $V(G) \setminus S$ into disjoint subsets A and B such that there are no edges between A and B and $|A|, |B| \leq \frac{2}{3}|V(G)|$.

Exercise 5.1. Check that the above definition of a separator is equivalent to requiring all connected components of $G \setminus S$ to have at most $\frac{2}{3}|V(G)|$ vertices.

Exercise 5.2. (a) Show that every tree has a one-vertex separator.

(b) We could also define a β-separator, for $\beta \in (0, 1)$, by replacing $\frac{2}{3}$ in the above definition by β. Check that for $\beta < \frac{2}{3}$, there are trees with no one-vertex β-separator. From this point of view, the value $\frac{2}{3}$ is natural. (For most applications, though, having β-separators for a constant $\beta < 1$ is sufficient, and the specific value of β is not too important.)

Separator theorems are results asserting that all graphs in a certain class have "small" separators (much smaller than the number of vertices). They have lots of applications, and in particular, they are the basis of many efficient divide-and-conquer algorithms.

Probably the most famous separator theorem, and arguably one of the nicest and most useful ones, is the **Lipton–Tarjan separator theorem for planar graphs**, asserting that *every planar graph on n vertices has a separator of size $O(\sqrt{n})$*.

Exercise 5.3. Show that the $m \times m$ square grid has no separator of size $m/4$. Thus, the order of magnitude in the Lipton–Tarjan theorem cannot be improved.

The separator theorem has several proofs (let us mention a simple graph-theoretic proof by Alon et al. [3] and a neat proof from the Koebe–Andreev–Thurston theorem mentioned in Section 1; see, e.g., [24]). There are scores of generalizations and variations. For example, every class of graphs with a fixed excluded minor admits $O(\sqrt{n})$-size separators [3].

Here we focus on a separator theorem for string graphs. Of course, in this case we cannot bound the separator size by a sublinear function of the number of vertices (because K_n is a string graph); for this reason, the bound is in terms of the number of *edges*.

Theorem 5.4. *Every string graph with $m \geq 2$ edges has a separator with $O(\sqrt{m} \log m)$ vertices.*

The first separator theorem for string graphs, with a worse bound of $O(m^{3/4} \sqrt{\log m})$, was proved by Fox and Pach [13]. The improvement to $O(\sqrt{m} \log m)$ was obtained while preparing this text, and it was published in a concise form in [23].

The proof, whose exposition will occupy most of the rest of this chapter, is a remarkable chain of diverse ideas coming from various sources.

Fox and Pach conjectured that the theorem should hold with $O(\sqrt{m})$. This, if true, would be asymptotically optimal: we already know this for graphs with n vertices and $O(n)$ edges, since every planar graph is a string graph, but asymptotic optimality holds for graphs with any number of edges between n and $\binom{n}{2}$.

The separator theorem shows that string graphs are (globally) very different from typical (not too dense) random graphs and, more generally, from expanders.

In the next section, we demonstrate a surprising use of Theorem 5.4; for a number of other applications we refer to [13, 14]. Then we start working on the proof of Theorem 5.4 in Section 7.

6   Crossing number versus pair-crossing number

Now something else: we will discuss crossing numbers in this section. The **crossing number** cr(G) of a graph G is the smallest possible number of edge intersections (crossings) in a standard drawing of G. We recall that in a standard drawing, edges are polygonal lines with finitely many intersections and no triple points; see Section 3. In this section we consider only standard drawings.

One may also consider the **rectilinear crossing number** $\overline{\text{cr}}(G)$, which is the minimum number of crossings in a straight-edge drawing of G, but this behaves very differently from $\text{cr}(G)$, and the methods involved in its study are also different from those employed for the crossing number.

An algorithmic remark   The crossing number is an extensively studied and difficult graph parameter. Let us just mention in passing that computing $\text{cr}(G)$ is known to be NP-complete, but a tantalizing open problem is, how well it can be approximated (in polynomial time). On the one hand, there is a constant $c > 1$ such that $\text{cr}(G)$ is hard to approximate within a factor of c [6], and on the other hand, there is a (highly complicated) algorithm [9] with approximation factor roughly $n^{9/10}$ for n-vertex graphs with maximum degree bounded by a constant. The latter result may not look very impressive, but it is the first one breaking a long-standing barrier of n.

A difficult case here are graphs with relatively small, but not too small, crossing number, say around n. Indeed, on the one hand, for every fixed k, there is a linear-time algorithm deciding whether $\text{cr}(G) \leq k$ [15,18]. On the other hand, for a graph with maximum degree bounded by a constant and with crossing number k, a drawing with at most $O((n + k)(\log n)^2)$ crossings can be found in polynomial time; this is based on [22], with improvements of [2,11] (also see [10]).

The single-crossing lemma   Here is a useful basic fact about drawings minimizing the crossing number.

Lemma 6.1 (Single-crossing lemma). *In every (standard) drawing of G that minimizes the crossing number, no two edges intersect more than once.*

Proof. We show that if edges e, e' intersect at least twice, the number of crossings can be reduced. We consider crossings X_1 and X_2 that are consecutive along e. There are two cases to consider, the second one being easy to overlook, and in each of them we redraw the edges locally as indicated:

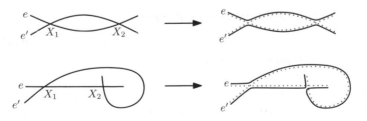

This reduces the total number of crossings by at least 2. We also note that since the part of e' between X_1 and X_2 may be intersected by e, this redrawing may introduce self-intersections of e—but these are easily eliminated by shortcutting the resulting loops on e.   □

Exercise 6.2. Let us say that X_1, X_2 is a *simple pair of crossings* of edges e and e' if X_1 and X_2 are consecutive on both e and e'. Draw an example of two edges that cross several times but that have no simple pair of crossings.

The lemma just proved shows that it does not matter whether we define the crossing number as the minimum number of *crossings* or as the minimum number of *crossing pairs of edges* in a drawing. Do you believe the previous sentence? If yes, you are in a good company, since many people got caught, and only Pach and Tóth [26], and independently Mohar (at a 1995 AMS Conference on Topological Graph Theory), noticed that the lemma proves nothing like that, since it is not clear whether a drawing with minimum number of crossings also has a minimum number of crossing pairs. Indeed, the number of crossing pairs may very well increase in a redrawing as in the proof of the lemma:

Thus, it makes good sense to define the **pair-crossing number** $\mathrm{pcr}(G)$ as the minimum possible number of pairs of edges that cross in a drawing of G. Clearly, $\mathrm{pcr}(G) \leq \mathrm{cr}(G)$ for all G.

It is generally conjectured that $\mathrm{pcr}(G) = \mathrm{cr}(G)$ for all G, but if true, this is unlikely to be proved by a "local" redrawing argument in the spirit of Lemma 6.1—at least, many people tried that and failed.

A warning example is a result of Pelsmajer et al. [25], who found a graph G with $\mathrm{ocr}(G) < \mathrm{cr}(G)$. Here $\mathrm{ocr}(G)$ is the **odd crossing number** of G, which is the minimum number, over drawings of G, of pairs of edges that cross *an odd number of times*. A good motivation for studying the odd crossing number is the famous **Hanani–Tutte theorem**, asserting that if a graph has a drawing in which every two *non-adjacent* edges cross an even number of times, then it is planar (see [27] for a modern treatment). In particular, this implies that $\mathrm{ocr}(G) = 0$, $\mathrm{pcr}(G) = 0$, and $\mathrm{cr}(G) = 0$ are all equivalent.

One direction of investigating the pcr/cr puzzle is to bound the crossing number by some function of the pair-crossing number, and to try to get as small a bound as possible. We begin with a simple result in this direction.

Proposition 6.3. *If* $\mathrm{pcr}(G) = k$, *then* $\mathrm{cr}(G) \leq 2k^2$.

Proof. We fix a drawing D of $G = (V, E)$ witnessing $\mathrm{pcr}(G)$. Let $F \subseteq E$ be the set of edges that participate in at least one crossing, and let $E_0 = E \setminus F$; thus, the edges of E_0 define a plane subgraph in D. We keep the drawing of these edges and we redraw the edges of F so that every two of them have at most one crossing, as in the proof of Lemma 6.1 (we note that E_0 does not interfere with this redrawing in any way). Since $|F| \leq 2k$, the resulting drawing has at most $\binom{2k}{2} \leq 2k^2$ crossings. □

The current strongest bound is based on the separator theorem for string graphs.

Theorem 6.4. *If* $\mathrm{pcr}(G) = k \geq 2$, *then* $\mathrm{cr}(G) = O(k^{3/2}(\log k)^2)$.

The following proof is due to Tóth [30]; he states a worse bound, but the bound above follows immediately from his proof by plugging in a better separator theorem.

We begin the proof with a variant of the single-crossing lemma (Lemma 6.1).

Lemma 6.5 (Red-blue single-crossing lemma). *Let* G *be a graph in which each edge is either red or blue, and let* D *be a drawing of* G. *Then there is a drawing* D' *of* G *such that the following hold:*

(i) *Each edge in* D' *is drawn in an arbitrarily small neighborhood of the edges of* D.
(ii) *Every two edges intersect at most once in* D'.
(iii) *The number of blue-blue crossings in* D' *is no larger than in* D.

Proof. While edges e, e' crossing at least twice exist, we repeat redrawing operations similar to those in Lemma 6.1. However, while in that lemma we swapped portions of e and e', here we keep e fixed and route e' along it,

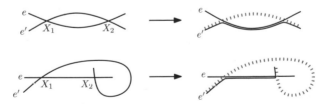

or the other way round. We may again introduce self-intersections of e or e', but we remove them by shortcutting loops.

To decide which way of redrawing to use, we let $\hat{e}$ be the portion of e between X_1 and X_2 (excluding X_1 and X_2), and similarly for $\hat{e}'$. Let b and

b' be the number of crossings of blue edges with $\hat{e}$ and $\hat{e}'$, respectively, and similarly for r, r'. If the pair (b, r) is lexicographically smaller than (b', r'), we route $\hat{e}'$ along $\hat{e}$, and otherwise, $\hat{e}$ along $\hat{e}'$.

It is clear that *if* this redrawing procedure terminates, then it yields a drawing D' satisfying (i) and (ii). To show that it terminates, and that (iii) holds, it suffices to check that each redrawing strictly lexicographically decreases the vector (x_{BB}, x_{RB}, x_{RR}) for the current drawing, where x_{BB} is the total number of blue-blue crossings, and similarly for x_{RB} and x_{RR}.

To check this, we distinguish three cases. If both e, e' are blue, then the redrawing decreases x_{BB}. If both e, e' are red, then x_{BB} stays the same and either x_{RB} decreases, or it stays the same and x_{RR} decreases. Finally, if e is blue and e' is red, then either x_{BB} decreases, or it stays the same and x_{RB} decreases. $\qquad\square$

In order to prove Theorem 6.4, we will proceed by induction. The inductive hypothesis is the following strengthening of the theorem.

Claim 6.6. Let $G = (V, E)$ be a graph, and let D be a drawing of G with k crossing pairs of edges and with $\ell \geq 2$ edges that have at least one crossing (thus, $\ell \leq 2k$ and $k \leq \binom{\ell}{2}$). Then there is a drawing D' of G such that every edge is drawn in an arbitrarily small neighborhood of the edges in D, and D' has at most $Ak^{3/2}(\log \ell)^2$ crossings, where A is a suitable constant.

Proof. We proceed by induction on ℓ (the proof will also establish the base case $\ell = 2$ directly).

As in the proof of Proposition 6.3, we first partition the edge set E into F, the edges with crossings, and $E_0 = E \setminus F$. Thus, $|F| = \ell$.

Let us consider the edges of F in the drawing D as strings (where we cut off tiny pieces near the vertices, so that the strings meet only if the corresponding edges cross). This defines a string graph with ℓ vertices and k edges.

By Theorem 5.4, this string graph has a separator of size at most $C\sqrt{k} \log k$, with a suitable constant C. This defines a partition of F into disjoint subsets F_0, F_1, F_1; we have $\ell_0 := |F_0| \leq C\sqrt{k} \log k, |F_1|, |F_2| \leq \frac{2}{3}\ell$, and no edge of F_1 crosses any edge of F_2. Let k_i be the number of crossing pairs of edges of F_i in $D, i = 1, 2$; we have $k_1 + k_2 \leq k$. Let ℓ_i be the number of edges of F_i that cross some edge of F_i.

Actually, Theorem 5.4 can be applied only if $k \geq 2$, while in our case, for $\ell = 2$ it may happen that $k = 1$. But if $k = 1$, we simply put $F_0 = F$ and $F_1 = F_2 = \emptyset$, and proceed with the subsequent argument.

Next, we apply the inductive hypothesis to the graphs $G_1 := (V, F_1)$ and $G_2 := (V, F_2)$ (drawn as in D). This yields drawings D_1' and D_2' as in the claim. (If F_1 has no crossings then, strictly speaking, the inductive hypothesis cannot be applied, but then D_1' can be taken as the plane drawing of G_1 inherited from D, and similarly for F_2.)

For $\ell_i \geq 2$, we can bound the number of crossings in D_i' by $Ak_i^{3/2}(\log \ell_i)^2$ according to the inductive hypothesis; for $\ell_i = 0$ there are no crossings (and $\ell_i = 1$ is impossible). We have $\ell_i \leq |F_i| \leq \frac{2}{3}\ell$, and hence $\log \ell_i \leq \log \ell - c_0$, where $c_0 = \log \frac{3}{2} > 0$. For the subsequent computation, it will be convenient to bound $(\log \ell_i)^2$ from above in a slightly strange-looking way: by $(\log \ell)(\log \ell - c_0)$. The resulting bound $Ak_i^{3/2}(\log \ell)(\log \ell - c_0)$ gives 0 for $k_i = 0$ and so it is also valid in the case $k_i = \ell_i = 0$.

We overlay D_1' and D_2' and add the edges of F_0 drawn as in D; this gives a drawing $\tilde{D}$ of the graph (V, F). Let us color the edges of $F_1 \cup F_2$ blue and those of F_0 red. By the above, and using $k_1^{3/2} + k_2^{3/2} \leq k^{3/2}$, the total number of blue-blue crossings in $\tilde{D}$ can be bounded by

$$A(k_1^{3/2} + k_2^{3/2})(\log \ell)(\log \ell - c_0) \leq Ak^{3/2}(\log \ell)^2 - Ac_0 k^{3/2} \log \ell.$$

The first term of the last expression is the desired bound for the number of all crossings, and thus the second term is the "breathing room" — we need to bound the number of red-blue and red-red crossings by $Ac_0 k^{3/2} \log \ell$.

We cannot control the number of red-blue and red-red crossings in $\tilde{D}$, but we apply Lemma 6.5 to the graph (V, F) with the drawing $\tilde{D}$. This provides a new drawing of (V, F), to which we add the edges of E_0 from the original drawing D, and this yields the final drawing D'. The number of blue-blue crossings in D' is no larger than in $\tilde{D}$, and the number of red-blue and red-red crossings is at most $|F_0| \cdot |F| = \ell_0 \ell$. Using $\ell \leq 2k$ and $k \leq \binom{\ell}{2} \leq \ell^2$, we further bound this by $(C\sqrt{k} \log k)2k \leq 4Ck^{3/2} \log \ell \leq Ac_0 k^{3/2} \log \ell$, provided that A was chosen sufficiently large.

This concludes the inductive proof of the claim and thus yields Theorem 6.4. $\square$

7 Multicommodity flows, congestion, and cuts

We start working towards a proof of the separator theorem for string graphs (Theorem 5.4). The overall scheme of the proof is given at the end of this section, but most of the actual work remains for later sections.

s-t flows   As a motivation for the subsequent developments, we briefly recall the duality between flows and cuts in graphs. If $G = (V, E)$ is a graph with two distinguished vertices s and t, in which every edge has unit capacity, then the maximum amount of flow from s to t equals the minimum number of edges we have to remove in order to destroy all *s-t* paths. There is also a more general weighted version, in which the capacity of each edge e is a given real number $w_e \geq 0$.

Multicommodity flows   Instead of flows between just two vertices, we will use *multicommodity flows*; namely, we want a unit flow between every pair $\{u, v\}$ of vertices of the considered graph.

Let us remark that instead of requiring unit flow for every pair, we can consider an arbitrary *demand function* $D\binom{V}{2} \rightarrow [0, \infty)$, specifying some demand $D(u, v)$ on the flow between u and v for every pair $\{u, v\}$. All of the considerations below can be done in this more general setting; we can also put weights on edges and vertices. For simplicity, we stick to the unweighted case, which is sufficient for us; conceptually, the weighted case mostly brings nothing new.

For our purposes, it is convenient to formalize a multicommodity flow as an assignment of nonnegative numbers to paths in the considered graph. Thus, we define $\mathcal{P}$ to be the set of all paths (of nonzero length) in G, and a **multicommodity flow** is a mapping $\varphi \, \mathcal{P} \rightarrow [0, \infty)$. Since we will talk almost exclusively about multicommodity flows, we will sometimes say just "flow" instead of "multicommodity flow".

The amount of flow between two vertices u, v is $\sum_{P \in \mathcal{P}_{uv}} \varphi(P)$, where $\mathcal{P}_{uv} \subseteq \mathcal{P}$ is the set of all paths with end-vertices u and v. If we think of the vertices as cities and the edges as roads, then $\varphi(P), P \in \mathcal{P}_{uv}$ may be the number of cars per hour driving from u to v or from v to u along the route P.

Edge congestion   The requirement of unit flow between every pair of vertices is expressed as $\sum_{P \in \mathcal{P}_{uv}} \varphi(P) \geq 1, \{u, v\} \in \binom{V}{2}$. We define the **edge congestion** under φ as

$$\mathrm{econg}(\varphi) = \max_{e \in E} \sum_{P \in \mathcal{P}: e \in P} \varphi(P),$$

and the edge congestion of G as $\mathrm{econg}(G) = \min_\varphi \mathrm{econg}(\varphi)$, where the minimum is over all flows with unit flow between every two vertices.[2] If G is disconnected, then there are no flows φ as above, and we have $\mathrm{econg}(G) = \infty$.

Let us consider an **edge cut** in G, which for us is a partition $(A, V \setminus A)$ of V into two *nonempty* subsets. By $E(A, V \setminus A)$ we denote the set of all edges of G connecting A to $V \setminus A$. If there is a unit flow between every two vertices of G, then $|A| \cdot |V \setminus A|$ units of flow have to pass through

---

[2] A compactness argument, which we omit, shows that the minimum is actually attained.

the edges of $E(A, V \setminus A)$, and hence

$$\text{econg}(G) \geq \frac{|A| \cdot |V \setminus A|}{|E(A, V \setminus A)|}.$$

If we define the *sparsity* of the edge cut $(A, V \setminus A)$ as

$$\text{espars}(A, V \setminus A) = \frac{|E(A, V \setminus A)|}{|A| \cdot |V \setminus A|},$$

and the **edge sparsity**[3] $\text{espars}(G) := \min_A \text{espars}(A, V \setminus A)$, we can write the conclusion of the previous consideration compactly as $\text{espars}(G) \geq 1/\text{econg}(G)$.

Approximate duality   Unlike in the case of s-t flows, it turns out that the last inequality can be strict.

Exercise 7.1. Let G be an n-vertex **constant-degree expander**, which means that, for some constants Δ and $\beta > 0$, all degrees in G are at most Δ and $\text{espars}(G) \geq \frac{\beta}{n}$. The existence of such graphs, with some Δ, β fixed and n arbitrarily large, is well known; see, e.g., [16]. Prove that $\text{econg}(G) > 1/\text{espars}(G)$ (assuming that n is sufficiently large in terms of Δ and β), and actually, that $\text{econg}(G) = \Omega(\frac{\log n}{\text{espars}(G)})$. Hint: show that, say, half of the vertex pairs have distance $\Omega(\log n)$.

However, an important result, discovered by Leighton and Rao [22], asserts that the gap between the two quantities cannot be very large; this is an instance of *approximate duality* between multicommodity flows and cuts.

Theorem 7.2 (Approximate duality, edge version). *For every n-vertex graph G we have*

$$\text{espars}(G) = O\left(\frac{\log n}{\text{econg}(G)}\right).$$

Although we won't really need this particular theorem, the proof can serve as an introduction to things we will actually use, and we present it in Section 9.

Exercise 7.3 (Edge sparsity and balanced edge cut). Let $\beta > 0$ and let G be a graph on n vertices such that $\text{espars}(H) \leq \beta$ for every induced subgraph H of G on at least $\frac{2}{3}n$ vertices. Show that G has a balanced edge cut $(A, V \setminus A)$ with $|E(A, V \setminus A)| \leq \beta n^2$, where balanced means that $\frac{1}{3}n \leq |A| \leq \frac{2}{3}n$.

---

[3] What we call edge sparsity is often called just *sparsity*. This quantity is also closely related to the *Cheeger constant*, or *edge expansion*, of G, which is defined as $\min_{A \subseteq V:1 \leq |A| \leq |V|/2}(|E(A, V \setminus A)|/|A|)$.

Vertex notions   For the proof of the separator theorem for string graphs, we will need vertex analogs of the "edge" notions and results just discussed.

For a flow φ in G we introduce the **vertex congestion**

$$\mathrm{vcong}(\varphi) := \max_{v \in V} \mathrm{vcong}(v), \quad \text{where} \quad \mathrm{vcong}(v) := \sum_{P \in \mathcal{P}: v \in P} \frac{1}{2} \varphi(P),$$

and the vertex congestion of G is $\mathrm{vcong}(G) := \min_\varphi \mathrm{vcong}(\varphi)$, where the minimum is over all φ with unit flow between every two vertices. The $\frac{1}{2}$ in the above formula should be interpreted as 1 if v is an inner vertex of the path P, and as $\frac{1}{2}$ if v is one of the end-vertices of P. We thus think of the flow along P as incurring congestion $\frac{1}{2}\varphi(P)$ when entering a vertex and congestion $\frac{1}{2}\varphi(P)$ when leaving it. (This convention is a bit of a nuisance in the definition of vertex congestion, but later on, it will pay off when we pass to a "dual" notion.)

By a **vertex cut** in G we mean a partition (A, B, S) of V into three disjoint subsets such that $A \neq \emptyset \neq B$ and there are no edges between A and B (this is like in the definition of a separator, except that we do not require the sizes of A and B to be roughly the same).

If φ sends unit flow between every pair of vertices, then the flows between A and B contribute a total flow of $|A| \cdot |B|$ through S, and moreover, from each vertex of S we have a flow of $n - 1$ to the remaining vertices. Thus the total congestion of the vertices in S is at least $|A| \cdot |B| + \frac{1}{2}|S|(n - 1)$. Losing a constant factor (and using $n \geq 2$), we bound this somewhat unwieldy expression from below by $\frac{1}{4}|A| \cdot |B| + \frac{1}{4}|S|n = \frac{1}{4}|A \cup S| \cdot |B \cup S|$.

This suggests to define the *sparsity* of a vertex cut (A, B, S) as

$$\mathrm{vspars}(A, B, S) := \frac{|S|}{|A \cup S| \cdot |B \cup S|},$$

and the **vertex sparsity** of G is $\mathrm{vspars}(G) := \min_{(A,B,S)} \mathrm{vspars}(A, B, S)$, where the minimum is over all vertex cuts.

By the considerations above, we have $\mathrm{vcong}(G) \geq 1/(4\,\mathrm{vspars}(G))$. We will need the following analog of Theorem 7.2:

Theorem 7.4 (Approximate duality, vertex version).   *For every connected n-vertex graph G we have*

$$\mathrm{vspars}(G) = O\!\left(\frac{\log n}{\mathrm{vcong}(G)}\right).$$

The proof is deferred to Section 11.

Exercise 7.5 (Vertex sparsity and separators). Let $\alpha > 0$ and let G be a graph on n vertices such that $\mathrm{vspars}(H) \leq \alpha$ for every induced subgraph H of G on at least $\frac{2}{3}n$ vertices. Show that G has a separator of size at most αn^2.

String graphs have large vertex congestion   The last ingredient in the proof of the separator theorem for string graphs is the following result about string graphs.

Proposition 7.6. *For every connected string graph G with n vertices and m edges, we have*

$$\frac{1}{\mathrm{vcong}(G)} = O\left(\frac{\sqrt{m}}{n^2}\right).$$

This is the only specific property of string graphs used in the proof of the separator theorem. The next section is devoted to the proof of this proposition.

Proof of the separator theorem for string graphs (Theorem 5.4).
Let G be a string graph with n vertices and $m \geq n$ edges. We have $1/\mathrm{vcong}(G) = O(\sqrt{m}/n^2)$ by Proposition 7.6. By approximate duality (Theorem 7.4), we have $\mathrm{vspars}(G) = O((\log n)\sqrt{m}/n^2)$, and so G has a separator of size $O(\sqrt{m}\,\log n)$ according to Exercise 7.5.   □

Exercise 7.7. Let G be a string graph with m edges whose maximum degree is bounded by a constant Δ. Derive from Theorem 7.2 (the edge version of the approximate duality) and from Proposition 7.6 that G has a separator of size $O(\sqrt{m}\,\log m)$, where the implicit constant may depend on Δ.

8   String graphs have large vertex congestion

The strategy of the proof of Proposition 7.6 is this: Given a string representation of an n-vertex graph G and a multicommodity flow in G with a small vertex congestion, we will construct a drawing of K_n in which only a small number of edge pairs cross. This will contradict the following result:

Lemma 8.1. *For $n \geq 5$, $\mathrm{pcr}(K_n) = \Omega(n^4)$.*

The proof of this lemma relies on the following fact.

Fact 8.2. In every plane drawing of K_5, some two *independent* edges intersect, where independent means that the edges do not share a vertex.

This fact is a consequence of the Hanani–Tutte theorem mentioned above Proposition 6.3, although that theorem is somewhat too big a hammer for this purpose. But proving the fact rigorously is harder than it may seem, even if we assume nonplanarity of K_5 as known (although a rigorous proof of the nonplanarity is almost never included in graph theory courses).

Exercise 8.3. Find a mistake in the following "proof" of Fact 8.2: Consider a (standard) drawing of K_5. If two independent edges cross, we are done, and otherwise, some two edges sharing a vertex cross. But such crossings can be removed by the following transformation

and so eventually we reach a plane drawing—a contradiction.

Proof of Lemma 8.1. By Fact 8.2, in every drawing of K_n, every 5-tuple of vertices induces a pair of independent edges that cross. A given pair of independent crossing edges determines 4 vertices of the 5-tuple, and so the number of 5-tuples inducing this particular pair of edges is at most $n - 4$. So $\mathrm{pcr}(K_n) \geq \binom{n}{5}/(n - 4) = \Omega(n^4)$.   □

We remark that Lemma 8.1 also follows from a generally useful result, the **crossing lemma** of Ajtai et al. [1] and Leighton [20], which asserts that every graph with n vertices and $m \geq 4n$ edges has crossing number $\Omega(m^3/n^2)$. We actually need a version of the lemma for the pair-crossing number, which holds with the same bound, as was observed in Pach and Tóth [26, Thm. 3].[4] This proof does not avoid Fact 8.2—it actually relies on a generalization of it.

Proof of Proposition 7.6. Let $G = (V, E)$, and let $(\gamma_v : v \in V)$ be a string representation of G. We are going to produce a drawing of the complete graph K_V on the vertex set V.

We draw each vertex $v \in V$ as a point $p_v \in \gamma_v$, in such a way that all the p_v are distinct.

For every edge $\{u, v\} \in \binom{V}{2}$ of the complete graph, we pick a path P_{uv} from $\mathcal{P}_{uv}$, in a way to be specified later. Let us enumerate the vertices along P_{uv} as $v_0 = u, v_1, v_2, \ldots, v_k = v$. Then we draw the edge $\{u, v\}$

---

[4] The argument in their proof is not quite correct, but the problem is rectified in Remark 2 in Section 3 of [26].

of K_V in the following manner: we start at p_u, follow γ_u until some (arbitrarily chosen) intersection with γ_{v_1}, then we follow γ_{v_1} until some intersection with γ_{v_2}, etc., until we reach γ_v and p_v on it.

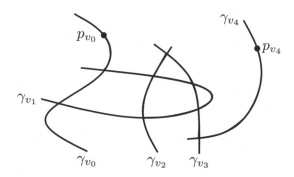

In this way we typically do not get a standard drawing, since edges may share segments, have self-intersections and triple points, and they may pass through vertices. However, we can obtain a standard drawing by shortcutting loops and a small perturbation of the edges, in such a way that no new intersecting pairs of edges are created. Hence, by Lemma 8.1, there are $\Omega(n^4)$ intersecting pairs of edges in the original drawing as well. We are going to estimate the number of intersecting pairs in a different way.

We note that the drawings of two edges $\{u, v\}$ and $\{u', v'\}$ cannot intersect unless there are vertices $w \in P_{uv}$ and $w' \in P_{u'v'}$ such that $\gamma_w \cap \gamma_{w'} \neq \emptyset$, i.e., $\{w, w'\} \in E(G)$ or $w = w'$. Let us write this latter condition as $P_{uv} \sim P_{u'v'}$.

How do we select the paths P_{uv}? For this, we consider a flow φ for which vcong(G) is attained. Since there is a unit flow between every pair of vertices $\{u, v\}$, the values of $\varphi(P)$ define a probability distribution on $\mathcal{P}_{uv}$. We choose $P_{uv} \in \mathcal{P}_{uv}$ from this distribution at random, the choices independent for different $\{u, v\}$.

The number X of intersecting pairs of edges in this drawing is a random variable, and we bound from above its expectation. First we note that

$$\text{Prob}\left[\{u, v\} \text{ and } \{u', v'\} \text{ intersect}\right] \leq \text{Prob}\left[P_{uv} \sim P_{u'v'}\right]$$

$$= \sum_{P \in \mathcal{P}_{uv}, P' \in \mathcal{P}_{u'v'}, P \sim P'} \text{Prob}\left[P_{uv} = P \text{ and } P_{u'v'} = P'\right]$$

$$= \sum_{P \in \mathcal{P}_{uv}, P' \in \mathcal{P}_{u'v'}, P \sim P'} \varphi(P)\varphi(P')$$

(for the last equality we have used independence). Therefore

$$
\begin{aligned}
\mathbf{E}[X] &= \sum_{\{\{u,v\},\{u',v'\}\}\in\binom{\binom{V}{2}}{2}} \mathrm{Prob}\big[\,\{u,v\}\text{ and }\{u',v'\}\text{ intersect}\big] \\
&\le \sum_{\{u,v\},\{u',v'\}} \sum_{P\in\mathcal{P}_{uv},\,P'\in\mathcal{P}_{u'v'},\,P\sim P'} \varphi(P)\varphi(P') \\
&= \sum_{\{w,w'\}\in E \text{ or } w=w'} \sum_{P,P'\in\mathcal{P},\,w\in P,\,w'\in P'} \varphi(P)\varphi(P') \\
&= \sum_{\{w,w'\}\in E \text{ or } w=w'} \Big(\sum_{P\in\mathcal{P},\,w\in P}\varphi(P)\Big)\Big(\sum_{P'\in\mathcal{P},\,w'\in P'}\varphi(P')\Big).
\end{aligned}
$$

The first sum in parentheses is at most $2\,\mathrm{vcong}(w)$, and the second one at most $2\,\mathrm{vcong}(w')$; the 2 is needed because of the paths P for which w is an end-vertex. The number of terms in the outer sum is $|E|+n\le 2m$. Altogether we get $\mathbf{E}[X]\le 8m\,\mathrm{vcong}(G)^2$.

Since, on the other hand, we always have $X=\Omega(n^4)$, we obtain $1/\mathrm{vcong}(G)=O(\sqrt{m}/n^2)$ as claimed. $\square$

9   Flows, cuts, and metrics: the edge case

Here we prove the edge version of the approximate flow/cut duality, Theorem 7.2. We essentially follow an argument of Linial, London, and Rabinovich [21].

Dualizing the linear program   The first step of the proof can be concisely expressed as follows: express $\mathrm{econg}(G)$ by a linear program, dualize it, and see what the dual means.[5]

It is slightly nicer to work with $\frac{1}{\mathrm{econg}(G)}$, which can be expressed as the maximum t such that there is a flow φ with edge congestion at most 1 that sends at least t between every pair of vertices. The resulting linear

---

[5] We assume that the reader has heard about linear programming and the duality theorem in it; if not, we recommend consulting a suitable source. Here is a very brief summary: A **linear program** is the computational problem of maximizing (or minimizing) a linear function over the intersection of finitely many half-spaces in $\mathbb{R}^n$ (*i.e.*, a convex polyhedron). Every linear program can be converted to a *standard form*: $\max_{x\in P} c^T x$ with $P=\{x\in\mathbb{R}^n : Ax\le b, x\ge 0\}$, where $c\in\mathbb{R}^n$, $b\in\mathbb{R}^m$, A is an $m\times n$ matrix, and the inequalities between vectors are meant componentwise. The *dual* of this linear program is $\min_{y\in D} b^T y$, $D=\{y\in\mathbb{R}^m, A^T y\ge c, y\ge 0\}$, and the *duality theorem of linear programming* asserts that if $P\ne\emptyset\ne D$, then $\min_{x\in P} c^T x=\max_{y\in D} b^T y$.

program has variables $t \geq 0$ and $\varphi(P)$, $P \in \mathcal{P}$, and it looks like this:

$$\max \left\{ t \geq 0 \; \varphi(P) \geq 0 \text{ for all } P \in \mathcal{P}, \right.$$

$$\sum_{P:e\in P} \varphi(P) \leq 1 \text{ for all } e \in E, \tag{9.1}$$

$$\left. \sum_{P\in\mathcal{P}_{uv}} \varphi(P) \geq t \text{ for all } \{u, v\} \in \binom{V}{2} \right\}. \tag{9.2}$$

The variables of the dual linear program are x_e, $e \in E$, corresponding to the constraints (9.1), and y_{uv}, $\{u, v\} \in \binom{V}{2}$, corresponding to the constraints (9.2). The dual reads

$$\min \left\{ \sum_{e\in E} x_e \; x_e, y_{uv} \geq 0, \right.$$

$$\sum_{e\in P} x_e \geq y_{uv}, \; P \in \mathcal{P}_{uv}, \; \{u, v\} \in \binom{V}{2}, \tag{9.3}$$

$$\left. \sum_{\{u,v\}\in\binom{V}{2}} y_{uv} \geq 1 \right\}, \tag{9.4}$$

and its value also equals $\frac{1}{\operatorname{econg}(G)}$ by the duality theorem. (Checking this claim carefully takes some work, and we expect only the most diligent readers to verify it—the others may simply take it for granted, since the linear programming duality is a side-topic for us.)

Fortunately, the dual linear program has a nice interpretation. We think of the variables x_e as edge weights, and then the constraints (9.3) say that y_{uv} is at most the sum of weights along every u-v path. From this it is easy to see that in an optimal solution of the dual linear program, each y_{uv} is the length of a shortest u-v path under the edge weights given by the x_e. When the y_{uv} are given in this way, we may also assume that for every edge $e = \{u, v\} \in E$, we have $x_e = y_{uv}$: Indeed, if $x_e > y_{uv}$, then there is a shortcut between u and v bypassing the edge e, i.e., a u-v path of length y_{uv}. So if we decrease x_e to the value y_{uv}, the length of a shortest path between every two vertices remains unchanged and thus no inequality in the linear program is violated, while $\sum_{e\in E} x_e$ decreases.

Thus, if we write d_w for the shortest-path (pseudo)metric[6] induced on V by an edge weight function $w \; E \to [0, \infty)$, we can express the con-

---

[6] We recall that a **metric** on a set V is a mapping $d \; V \times V \to [0, \infty)$ satisfying (i) $d(u, v) = d(v, u)$ for all u, v; (ii) $d(u, u) = 0$ for all u; (iii) $d(u, v) > 0$ whenever $u \neq v$; and (iv) $d(u, v) \leq d(u, x) + d(x, v)$ for all $u, v, x \in V$ (triangle inequality). A *pseudometric* satisfies the same axioms except possibly for (iii).

clusion of the dualization step as

$$\frac{1}{\mathrm{econg}(G)} = \min\left\{ \frac{\sum\limits_{\{u,v\}\in E} d_w(u,v)}{\sum\limits_{\{u,v\}\in\binom{V}{2}} d_w(u,v)} \ :\ w\ E \to [0,\infty),\ w \not\equiv 0 \right\}. \quad (9.5)$$

Here $w \not\equiv 0$ means that w is not identically 0; note that we replaced the constraint (9.4), requiring the sum of all distances under d_w to be at least 1, by dividing the minimized function by the sum of all distances.

Cut metrics and line metrics   To make further progress, we will investigate the minimum of the same ratio as in (9.5), but over different classes of metrics.

A **cut metric** on a set V is a pseudometric[7] c given by $c(u,v) = |f(u) - f(v)|$ for some function $f\ V \to \{0,1\}$.

By comparing the definitions, we can express the edge sparsity of a graph as

$$\mathrm{espars}(G) = \min\left\{ \frac{\sum_{\{u,v\}\in E} c(u,v)}{\sum_{\{u,v\}\in\binom{V}{2}} c(u,v)} \ :\ c\ \text{a cut metric on } V, c \not\equiv 0 \right\} \quad (9.6)$$

(please check).

Next, it turns out that we can replace cut metrics by line metrics in (9.6) and the minimum stays the same. Here a **line metric** is a pseudometric ℓ such that $\ell(u,v) = |f(u) - f(v)|$ for some function $f\ V \to \mathbb{R}$. We leave the proof as an instructive exercise.

Exercise 9.1. Show that the minimum in

$$\min\left\{ \frac{\sum_{\{u,v\}\in E} \ell(u,v)}{\sum_{\{u,v\}\in\binom{V}{2}} \ell(u,v)} \ :\ \ell\ \text{a line metric on } V, \ell \not\equiv 0 \right\} \quad (9.7)$$

is attained by a cut metric, and hence it also equals $\mathrm{espars}(G)$. (Hint: show that if the function f defining a line metric ℓ attains at least three distinct values, then some value can be eliminated.)

A key result that allows us to compare the minimum (9.5) over all shortest-path metrics with the minimum (9.7) over all line metric follows from the work of Bourgain [5]. His main theorem was formulated differently, but his proof immediately yields the following formulation, which is the most convenient for our purposes.

---

[7] Cut *metric* is really a misnomer, since a cut metric is almost never a metric; we should speak of a *cut pseudometric*, but we conform to the usage in the literature. A similar remark applies to line metrics considered below.

Theorem 9.2. *Let V be an n-point set. For every (pseudo)metric d on V there exists a line metric ℓ on V satisfying*

(i) *(ℓ is below d)* $\ell(u, v) \le d(u, v)$ *for all* $u, v \in V$, *and*
(ii) *(the average distance not decreased too much)*

$$\sum\nolimits_{\{u,v\}\in\binom{V}{2}} \ell(u, v) \ge \frac{c}{\log n}\sum\nolimits_{\{u,v\}\in\binom{V}{2}} d(u, v),$$

for a constant $c > 0$.

For completeness, we demonstrate the main idea of the proof in Section 10 below.

Proof of Theorem 7.2. Let d^* be a shortest-path metric attaining the minimum in the expression (9.5) for $\frac{1}{\mathrm{econg}(G)}$. We apply Theorem 9.2 with $d = d^*$ and obtain a line metric ℓ^* satisfying (i), (ii) in the theorem. Then

$$\frac{1}{\mathrm{econg}(G)} = \frac{\displaystyle\sum_{\{u,v\}\in E} d^*(u, v)}{\displaystyle\sum_{\{u,v\}\in\binom{V}{2}} d^*(u, v)} \ge \frac{c}{\log n} \cdot \frac{\displaystyle\sum_{\{u,v\}\in E} \ell^*(u, v)}{\displaystyle\sum_{\{u,v\}\in\binom{V}{2}} \ell^*(u, v)}$$

$$\ge \frac{c}{\log n} \cdot \min\left\{ \frac{\displaystyle\sum_{\{u,v\}\in E} \ell(u, v)}{\displaystyle\sum_{\{u,v\}\in\binom{V}{2}} \ell(u, v)} : \ell \text{ a line metric on } V, \ell \not\equiv 0 \right\}$$

$$= \frac{c}{\log n} \cdot \mathrm{espars}(G). \qquad \square$$

10   Proof of a weaker version of Bourgain's theorem

Here we prove a version of Theorem 9.2 with $\log n$ replaced by $\log^2 n$; this weakening makes the proof simpler, while preserving the main ideas.

Providing a line metric ℓ satisfying condition (i), $\ell \le d$, is equivalent to providing a function $f : V \to \mathbb{R}$ that is **1-Lipschitz**, *i.e.*, satisfies $|f(u) - f(v)| \le d(u, v)$ for all $u, v \in V$.

A suitable f is chosen at random, in the following steps.

1. Let k be the smallest integer with $2^k \ge n$, *i.e.*, $k = \lceil \log_2 n \rceil$. Choose an index $j \in \{0, 1, \ldots, k\}$ uniformly at random, and set $p := 2^{-j}$.
2. Choose a random subset $A \subseteq V$, where each point $v \in V$ is included in A independently with probability p.

3. Define f by $f(u) := d(u, A) = \min_{a \in A} d(u, a)$.

A nice thing about this way of choosing f is that it is 1-Lipschitz for every $A \subseteq V$, as can be easily checked using the triangle inequality. So it remains to show that, with positive probability, the line metric induced by f satisfies a weaker version of condition (ii), *i.e.*, that it does not decrease the average distance too much.

We will actually prove that for every $u, v \in V, u \neq v$,

$$\text{Prob}\left[|f(u) - f(v)| \geq \tfrac{c_0}{\log n} \cdot d(u, v)\right] \geq \frac{c_0}{\log n}, \qquad (10.1)$$

where the probability is with respect to the random choice of f as above, and $c_0 > 0$ is a suitable constant. Assuming (10.1), passing to expectation, and summing over $\{u, v\} \in \binom{V}{2}$, we arrive at

$$\mathbf{E}\left[\sum_{\{u,v\} \in \binom{V}{2}} |f(u) - f(v)|\right] \geq \frac{c_0^2}{\log^2 n} \sum_{\{u,v\} \in \binom{V}{2}} d(u, v),$$

and hence at least one f satisfies (ii) with $\log^2 n$ instead of $\log n$. So we fix u, v and we aim at proving (10.1). Let us set $\Delta := d(u, v)/(2k - 1)$. We have $|f(u) - f(v)| = |d(u, A) - d(v, A)|$, and the latter expression is at least Δ provided that, for some $r \geq 0$, the set A intersects the (closed) r-ball around u and avoids the (open) $(r + \Delta)$ ball around v, or the other way round.

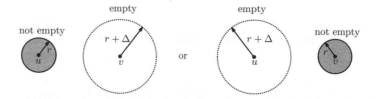

In order for this event to have a non-negligible probability, we need that the number of points in the bigger balls is not much larger than in the smaller ball. The trick for achieving this is to consider a system of balls as in the next picture:

The picture is for $k = 4$. In general, B_i is the closed ball of radius $i\Delta$, $i = 0, 1, \ldots, k$, centered at u for even i and at v for odd i. Let B_i° denote the corresponding open ball (all points at distance strictly smaller than $i\Delta$ from the center).

Let n_i be the number of points in B_i. We claim that $n_{i+1}/n_i \leq 2$ for some $i \in \{0, 1, \ldots, k - 1\}$; indeed, if not, then $|B_k| > 2^k \geq n - a$ contradiction.

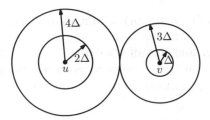

We fix such an i, and we also fix j_0 such that n_i is approximately 2^{j_0}; more precisely, $2^{j_0} \leq n_i < 2^{j_0+1}$.

Let $p = 2^{-j_0}$, and let us pick a random A as in the second step of the choice of f with this value of p. By a simple calculation, which we leave as an exercise, there is a constant $c_1 > 0$ such that

$$\text{Prob}\left[A \cap B_i \neq \emptyset \text{ and } A \cap B_{i+1}^{\circ} = \emptyset\right] \geq c_1.$$

Exercise 10.1. Let X, Y be disjoint sets, and let $A \subseteq X \cup Y$ be a random subset, where each point of $X \cup Y$ is included in A with probability p, independent of all other points, with $0 < p \leq \frac{1}{2}$. Assuming $\frac{1}{2p} \leq |X|, |Y| \leq \frac{2}{p}$, show that $\text{Prob}[A \cap X \neq \emptyset \text{ and } A \cap Y = \emptyset] \geq c_1$ for a constant $c_1 > 0$.

Now (10.1) follows easily: given u, v, the probability of choosing $j = j_0$ is $\frac{1}{k+1} = \Omega(\frac{1}{\log n})$, and conditioned on this choice, we have $\text{Prob}[|f(u) - f(v)| \geq \Delta] \geq c_1$. This concludes the proof of the weaker version of Theorem 9.2.

11 Flows, cuts, and metrics: the vertex case

Here we prove Theorem 7.4, the vertex case of the approximate duality, and this will also conclude the quest for the proof of the separator theorem for string graphs. Initially we proceed in a way similar to the edge case from Section 9, but the last step is more demanding and uses a nice method for producing sparse vertex cuts algorithmically.

Dualization again   As before, we write $\frac{1}{\text{vcong}(G)}$ as a linear program and dualize it. The linear program differs from the one for $\frac{1}{\text{econg}(G)}$ only in the second line:

$$\max \left\{ t \geq 0 \; \varphi(P) \geq 0 \text{ for all } P \in \mathcal{P}, \right.$$

$$\sum_{P: v \in P} \varphi(P) \leq 1 \text{ for all } v \in V, \qquad (11.1)$$

$$\left. \sum_{P \in \mathcal{P}_{uv}} \varphi(P) \geq t \text{ for all } \{u, v\} \in \binom{V}{2} \right\}; \qquad (11.2)$$

here the meaning of $\frac{1}{2}$ is as in the definition of vcong(G) in Section 7. In the dual, we have variables y_{uv} indexed by pairs of vertices and z_v indexed by vertices, and it reads

$$\min\left\{ \sum_{z\in V} z_v; \; z_v, \, y_{uv} \geq 0, \right.$$

$$\sum_{v\in P} \tfrac{1}{2} z_v \geq y_{uv}, \; P \in \mathcal{P}_{uv}, \; \{u,v\} \in \tbinom{V}{2}, \quad (11.3)$$

$$\left. \sum_{\{u,v\}\in\binom{V}{2}} y_{uv} \geq 1 \right\}. \quad (11.4)$$

This, too, can be interpreted using a metric on G. This time we have a function $s \; V \to [0,\infty)$ assigning weights to vertices. Let us define the derived weight of an edge $e = \{u,v\}$ by $w(e) := \tfrac{1}{2}(s(u)+s(v))$ and denote the corresponding shortest-path metric by d_s. Then, in a way very similar to the edge case, one can see that

$$\frac{1}{\text{vcong}(G)} = \min\left\{ \frac{\sum_{v\in V} s(v)}{\sum_{\{u,v\}\in\binom{V}{2}} d_s(u,v)} : s \; V \to [0,\infty), s \neq 0 \right\}. \quad (11.5)$$

Here the convention with $\frac{1}{2}$ for the vertex congestion pays off—the dual has a nice interpretation in terms of shortest-path metrics.

Let s^* be a weight function for which the minimum in (11.5) is attained. Applying Bourgain's theorem (Theorem 9.2) to the metric d_{s^*} yields a function $f^* \; V \to \mathbb{R}$ that is 1-Lipschitz w.r.t. d_{s^*} and satisfies

$$\frac{\sum_{v\in V} s(v)}{\sum_{\{u,v\}\in\binom{V}{2}} |f^*(u)-f^*(v)|} = O\left(\frac{\log n}{\text{vcong}(G)}\right).$$

The following theorem of Feige, Hajiaghayi, and Lee [12] then shows how such an f^* can be used to produce sparse vertex cuts in G. This is the last step in the proof of Theorem 7.4.

Theorem 11.1. *Let G be a graph, $s \; V \to [0,\infty)$ a weight function on the vertices, d_s the corresponding metric, and let $f \; V \to \mathbb{R}$ be a non-constant 1-Lipschitz function w.r.t. d_s. Then*

$$\text{vspars}(G) \leq \frac{\sum_{v\in V} s(v)}{\sum_{\{u,v\}\in\binom{V}{2}} |f(u)-f(v)|}.$$

Proof. The proof actually provides a polynomial-time algorithm for finding a vertex cut with sparsity bounded as in the theorem.

Let us number the vertices of G so that $f(v_1) \leq f(v_2) \leq \cdots \leq f(v_n)$. For every $i = 1, 2, \ldots, n-1$, we are going to find a vertex cut (A_i, B_i, S_i), and show that one of these will do.

To this end, given i, we form an auxiliary graph G_i^+ by adding new vertices x and y to G, connecting x to v_1 through v_i, and y to v_{i+1} through v_n.

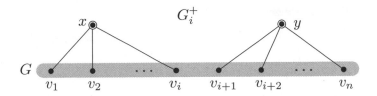

We let $S_i \subseteq V$ be a minimum cut in G_i^+ separating x from y (which can be found using a max-flow algorithm, for example). Let $A_i := \{v_1, \ldots, v_i\} \setminus S_i$ and $B_i := \{v_{i+1}, \ldots, v_n\} \setminus S_i$, and let

$$\alpha := \min_i \operatorname{vspars}(A_i, B_i, S_i) = \min_i \frac{|S_i|}{|A_i \cup S_i| \cdot |B_i \cup S_i|}.$$

Since $\{v_1, \ldots, v_i\} \subseteq A_i \cup S_i$, we have $|A_i \cup S_i| \geq i$, and similarly $|B_i \cup S_i| \geq n - i$. Thus, for every i we have

$$|S_i| \geq \alpha i (n - i). \tag{11.6}$$

In order to prove the theorem, we want to derive

$$\alpha \sum_{\{u,v\} \in \binom{V}{2}} |f(u) - f(v)| \leq \sum_{v \in V} s(v). \tag{11.7}$$

Setting $\varepsilon_i = f(v_{i+1}) - f(v_i) \geq 0$, we can rearrange the left-hand side: $\alpha \sum_{i<j}(f(v_j) - f(v_i)) = \alpha \sum_{i=1}^{n-1} i(n-i)\varepsilon_i$ (we just look how many times the segment between $f(v_i)$ and $f(v_{i+1})$ is counted). Then, substituting from (11.6), we finally bound the left-hand side of (11.7) by $\sum_{i=1}^{n-1} \varepsilon_i |S_i|$. It remains to prove

$$\sum_{i=1}^{n-1} \varepsilon_i |S_i| \leq \sum_{v \in V} s(v), \tag{11.8}$$

and this is the most ingenious part of the proof.

Roughly speaking, for every term $\varepsilon_i |S_i|$, we want to find vertices of sufficient total weight sufficiently close to the interval $[f(v_i), f(v_{i+1})]$. We use Menger's theorem, which guarantees that there are $|S_i|$ vertex-disjoint paths that have to "jump over" the interval $[f(v_i), f(v_{i+1})]$.

More precisely, we express both sides of (11.8) as integrals. Namely, we write $\sum_{i=1}^{n-1} \varepsilon_i |S_i| = \int_{-\infty}^{\infty} g(z)\,dz$, where g is the function that equals $|S_i|$ on $[f(v_i), f(v_{i+1}))$ and 0 elsewhere:

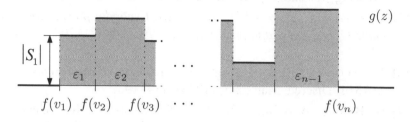

Similarly, $\sum_{v \in V} s(v) = \int_{-\infty}^{\infty} \sum_{i=1}^{n} h_i(z)\,dz$, where h_i is the function equal to 1 on $[f(v_i) - \frac{s(v_i)}{2}, f(v_i) + \frac{s(v_i)}{2}]$ and to 0 elsewhere:

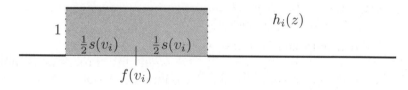

We claim that $g(z) \leq \sum_{i=1}^{n} h_i(z)$ for every $z \in \mathbb{R}$; this will imply (11.8).

Let $z \in [f(v_i), f(v_{i+1})]$, and set $m = g(z) = |S_i|$. We want to show $\sum_{i=1}^{n} h_i(z) \geq m$, which means that we need to find m distinct vertices v such that $|f(v) - z| \leq \frac{s(v)}{2}$; let us call such v the *paying vertices* since we can imagine that they pay for $g(z)$.

As announced, we use Menger's theorem, which tells us that, since S_i is a minimum x-y cut in G_i^+, there are m x-y paths $P_1, \ldots, P_m$ that are vertex-disjoint except for sharing the end-vertices x and y. Each P_j contains at least one edge $e_j = \{a_j, b_j\}$ with one endpoint among $v_1, \ldots, v_i$ and the other among $v_{i+1}, \ldots, v_n$. Hence $z \in [f(a_j), f(b_j)]$, and since f is 1-Lipschitz, $|f(a_j) - f(b_j)| \leq d_s(a_j, b_j) \leq \frac{1}{2}(s(a_j) + s(b_j))$. Thus, we have $|f(a_j) - z| \leq \frac{s(a_j)}{2}$ or $|f(b_j) - z| \leq \frac{s(b_j)}{2}$ (or both), and so a_j or b_j is a paying vertex. This gives the desired m distinct paying vertices. $\square$

ACKNOWLEDGEMENTS. I am very grateful to Rado Fulek, Vincent Kusters, Jan Kynčl, and Zuzana Safernová for proofreading, comments, and corrections. It was a pleasure to teach the courses together with Pavel Valtr in Prague and with Michael Hoffmann and Emo Welzl in Zurich, and to work with Jan Kratochvíl on questions in string graphs as well as on many other things. I also thank an anonymous referee for numerous useful remarks and suggestions.

References

[1] M. AJTAI, V. CHVÁTAL, M. M. NEWBORN and E. SZEMERÉDI, *Crossing-free subgraphs*, Ann. Discrete Math. **12** (1982), 9–12.

[2] S. ARORA, S. RAO and U. VAZIRANI, *Expander flows, geometric embeddings and graph partitioning*, J. ACM **56**(2) (2009), Art. 5, 37.

[3] N. ALON, P. SEYMOUR and R. THOMAS, *Planar separators*, SIAM J. Discrete Math. **2**(7) (1994), 184–193.

[4] C. BERGE, *Who killed the Duke of Densmore?* In: "Queneau, Raymond, Italo Calvino, et al., Oulipo Laboratory", Atlas Press, London, 1995.

[5] J. BOURGAIN, *On Lipschitz embedding of finite metric spaces in Hilbert space*, Israel J. Math. **52** (1985), 46–52.

[6] S. CABELLO, *Hardness of approximation for crossing number*, Discrete Comput. Geom. **49**(2) (2013), 348–358.

[7] J. CHALOPIN and D. GONÇALVES, *Every planar graph is the intersection graph of segments in the plane*, In: "Proc. 41st Annual ACM Symposium on Theory of Computing (STOC)", New York, NY, USA, 2009, ACM, 631–638.

[8] J. CHALOPIN, D. GONÇALVES and P. OCHEM, *Planar graphs have 1-string representations*, Discrete Comput. Geom. **43**(3) (2010), 626–647.

[9] J. CHUZHOY, *An algorithm for the graph crossing number problem*, In: "Proceedings of the 43rd annual ACM symposium on Theory of Computing (STOC)", (2011), 303–312. Full version in arXiv:1012.0255.

[10] J. CHUZHOY, Y. MAKARYCHEV and A. SIDIROPOULOS, *On graph crossing number and edge planarization*, In: "Proceedings of the Twenty-Second Annual ACM-SIAM Symposium on Discrete Algorithms", SIAM, Philadelphia, PA (2011), 1050–1069.

[11] G. EVEN, S. GUHA and B. SCHIEBER, *Improved approximations of crossings in graph drawings and VLSI layout areas*, SIAM J. Comput. **32**(1) (2002/2003), 231–252 (electronic).

[12] U. FEIGE, M. T. HAJIAGHAYI and J. R. LEE, *Improved approximation algorithms for minimum weight vertex separators*, SIAM Journal on Computing **38**(2) (2008), 629–657.

[13] J. FOX and J. PACH, *A separator theorem for string graphs and its applications*, Combinatorics, Probability & Computing **19**(3) (2010), 371–390.

[14] J. FOX and J. PACH, *Applications of a new separator theorem for string graphs*, Combinatorics, Probability and Computing **23** (2014), 66–74. Preprint arXiv:1302.7228.

[15] M. GROHE *Computing crossing numbers in quadratic time*, J. Comput. System Sci. **68**(2) (2004), 285–302.

[16] S. HOORY, N. LINIAL and A. WIDGERSON, *Expander graphs and their applications*, Bull. Am. Math. Soc., New Ser. **43** (4) (2006), 439–561.

[17] J. KRATOCHVÍL and J. MATOUŠEK, *String graphs requiring exponential representations*, J. Combin. Theory Ser. B **53**(1) (1991), 1–4.

[18] K. KAWARABAYASHI and B. REED, *Computing crossing number in linear time*, In: "Proc. 39th Annual ACM Symposium on Theory of Computing (STOC)", 2007, 382–390.

[19] J. KRATOCHVÍL, *String graphs. II: Recognizing string graphs is NP-hard*, J. Comb. Theory, Ser. B **52**(1) (1991), 67–78.

[20] F. T. LEIGHTON, *New lower bound techniques for VLSI*, Math. Systems Theory, **17** (1984), 47–70.

[21] N. LINIAL, E. LONDON and YU. RABINOVICH, *The geometry of graphs and some its algorithmic applications*, Combinatorica **15** (1995), 215–245.

[22] F. T. LEIGHTON and S. RAO, *Multicommodity max-flow min-cut theorems and their use in designing approximation algorithms*, J. Assoc. Comput. Machin. **46** (1999), 787–832.

[23] J. MATOUŠEK, *Near-optimal separators in string graphs*, Combinatorics, Probability and Computing **23** (2014), 135–139.

[24] J. PACH and P. K. AGARWAL, "Combinatorial Geometry" John Wiley & Sons, New York, NY, 1995.

[25] M. J. PELSMAJER, M. SCHAEFER and D. ŠTEFANKOVIČ, *Odd crossing number and crossing number are not the same* Discrete Comput. Geom. **39**(1-3) (2008), 442–454.

[26] J. PACH and G. TÓTH, *Which crossing number is it anyway?*, J. Combin. Theory Ser. B **80** (2000), 225–246.

[27] M. SCHAEFER, *Hanani–Tutte and related results*, In: "Geometry - Intuitive, Discrete, and Convex (Bólyai Society Mathematical Studies)", Vol. 24, I. Bárány et al. (eds.), Springer, Berlin, 2014.

[28] M. SCHAEFER and D. ŠTEFANKOVIČ, *Decidability of string graphs*, J. Comput. Syst. Sci. **68** (2004), 319–334. Preliminary version in Proc. 33rd Annual ACM Symposium on Theory of Computing, 2001.

[29] M. SCHAEFER, E. SEDGWICK and D. ŠTEFANKOVIČ, *Recognizing string graphs in NP*, J. Comput. Syst. Sci. **67**(2) (2003), 365–380.

[30] G. TÓTH, *A better bound for pair-crossing number*, In: J. Pach (ed.), "Thirty Essays on Geometric Graph Theory", Springer, Berlin, 2012, 563–567.

On first-order definable colorings

Jaroslav Nešetřil and Patrice Ossona de Mendez

Abstract. We address the problem of characterizing H-coloring problems that are first-order definable on a fixed class of relational structures. In this context, we give also several characterizations of a homomorphism dualities arising in a class of structures.

1 Introduction

Recall that *classical model theory* studies properties of abstract mathematical structures (finite or not) expressible in first-order logic [20], and *finite model theory* is the study of first-order logic (and its various extensions) on finite structures [11,27].

Constraint Satisfaction Problems (CSPs), and more specifically H-coloring problems, are standard examples of problems which can be expressed in monadic second order logic but usually not in the first-order logic. Of course, expressing a H-coloring problem in first-order logic would be highly appreciable, as it would allow fast checking (in at most polynomial time) although problems expressed in monadic second order logic are usually NP-complete. In this direction, it has been proved by Hell and Nešetřil [18] that in the context of finite undirected graphs the H-coloring problem is NP-complete unless H is bipartite, in which case the H-coloring problem is clearly polynomially solvable. This, and a similar dichotomy result of Schaefer [41], led Feder and Vardi [14, 15] to formulate the celebrated Dichotomy Conjecture which asserts that,

The first author was supported by grant ERCCZ LL-1201, CE-ITI P202/12/6061, and by the European Associated Laboratory "Structures in Combinatorics" (LEA STRUCO).

The second author was upported by grant ERCCZ LL-1201, by the European Associated Laboratory "Structures in Combinatorics" (LEA STRUCO), and partially supported by ANR project Stint under reference ANR-13-BS02-0007.

for every constraint language over an arbitrary finite domain, the corresponding constraint satisfaction problems are either solvable in polynomial time, or are NP-complete. It was soon noticed that this conjecture is equivalent to the existence of a dichotomy for (general) H-coloring problems, and in fact it suffices to prove it for oriented graphs (see [15] and [19]).

Alternatively, the class P of all polynomially solvable problems can be described as the class of problems expressible (on ordered structures) in first-order logic with a least fixed point operator [21, 44]. On the other hand the class NP may be characterized (up to polynomial equivalence) as the class of all problems which have a lift (or expansion) determined by forbidden homomorphisms from a finite set [25]. Hence, we are led naturally to the question of descriptive complexity of classes of structures corresponding to H-coloring problems. A particular case is the question whether a H-coloring problem may be expressed in first-order logic or not.

In this paper, we will consider the relativized version of the problem of first-order definability of H-coloring problems to graphs (or structures) belonging to a fixed class $\mathcal{C}$:

Problem 1.1. Given a fixed class $\mathcal{C}$ of graphs (directed graphs, relational structures), determine which H-coloring problems are first-order definable in $\mathcal{C}$. Explicitly, determine for which graphs (directed graphs, relational structures) H there exists a first-order sentence ϕ_H such that

$$\forall G \in \mathcal{C}: \qquad (G \models \phi_H) \iff (G \to H).$$

The case where $\mathcal{C}$ is the whole class of all finite graphs (all finite directed graphs, all finite relational structures with given finite signature) is well understood. Atserias [1,2] and Rossman [39] proved that in this case first-order definable H-colorings correspond exactly to *finite homomorphism dualities*, and these dualities have been fully characterized (for undirected graphs, by Nešetřil and Pultr [36]; for directed graphs, by Komárek [22]; for general finite structures, by Nešetřil and Tardif [37]) as follows:

Theorem 1.2 ([37]). *For any signature σ and any finite set $\mathcal{F}$ of σ-structures the following two statements are equivalent:*

1. *There exists D such that $\mathcal{F}$ and D form a finite duality, that is:*

$$\forall \text{ finite } G: \qquad (\forall F \in \mathcal{F}, F \nrightarrow G) \iff (G \to D)$$

2. $\mathcal{F}$ *is homomorphically equivalent to a set of finite (relational) trees.*

Note that an example of such a duality for the class of all finite directed graphs is the Gallai-Hasse-Roy-Vitaver theorem [16, 17, 40, 45], which states that for every directed graph $\vec{G}$ it holds:

$$\vec{P}_{k+1} \nrightarrow \vec{G} \quad \Longleftrightarrow \quad \vec{G} \rightarrow \vec{T}_k.$$

For general classes of graphs the answer is more complicated. For instance, let $\mathcal{C}$ be the class of (undirected) toroidal graphs and let ϕ be the sentence

$$\forall x_0 \forall x_1 \dots \forall x_{10} \bigvee_{i=0}^{10} \neg(x_i \sim x_{i+1}) \vee \neg(x_i \sim x_{i+2}) \vee \neg(x_i \sim x_{i+3}),$$

where additions are considered modulo 11 and where $u \sim v$ denotes that u and v are adjacent. Then, it follows from [43] that a graph $G \in \mathcal{C}$ satisfies ϕ if and only if it is 5-colourable. This property can be alternatively be expressed by the following *restricted duality*:

$$\forall G \in \mathcal{C}: \qquad \nrightarrow G \Longleftrightarrow G \longrightarrow$$

In fact the class of toroidal graphs has all restricted dualities in the following sense: for every connected F there exists H_F such that $F \nrightarrow H_F$ and for every toroidal graph G holds

$$F \nrightarrow G \quad \Longleftrightarrow \quad G \longrightarrow H_F.$$

For a general class of graphs $\mathcal{C}$, Problem 1.1 is very complex. We have to specialize. Hence we first require that the studied class $\mathcal{C}$ has some basic properties: we assume that

- $\mathcal{C}$ is *hereditary* (meaning that every induced subgraph of a graph in $\mathcal{C}$ is in $\mathcal{C}$);
- $\mathcal{C}$ is *addable* (meaning that disjoint unions of graphs in $\mathcal{C}$ are in $\mathcal{C}$);
- $\mathcal{C}$ is *topologically closed* (meaning that every subdivision of a graph in $\mathcal{C}$ is in $\mathcal{C}$).

We approach the problem of characterizing first-order definable colorings by first discriminating between the cases of sparse and dense classes of graphs using our *class taxonomy* [32, 33, 35], which seems relevant here. The second central ingredient of our study is the notion of *homomorphism preservation theorem* (HPT) for a class $\mathcal{C}$, which was investigated in [5, 7, 8, 35, 39]. Our approach can be outlined as follows:

If there exists some integer p such that every p-subdivision of a complete graph appears as a subgraph of some graph in $\mathcal{C}$ (meaning that $\mathcal{C}$ is *somewhere dense*) then we prove that every first-order definable H-coloring defines a restricted duality on the subclass $\mathcal{C}' \subseteq \mathcal{C}$ of the p-subdivisions of simple graphs (follows from HPT for $\mathcal{C}'$). Using a classical construction of Erdős [12], we deduce that if a H-coloring problem is first-order definable on $\mathcal{C}$ then either H is bipartite, or the odd-girth of H is at most $2p + 1$.

Otherwise (meaning that $\mathcal{C}$ is *nowhere dense*) it follows from HPT for nowhere dense classes that every first-order definable H-coloring defines a restricted duality on $\mathcal{C}$. In the case where there exists some integer p such that p-subdivisions of graphs with arbitrarily large average degree appear as subgraphs of graphs in $\mathcal{C}$ (meaning that $\mathcal{C}$ is not a *bounded expansion* class) we prove that H cannot be a restricted dual of a non-bipartite graph with arbitrarily large odd-girth. Modulo some reasonable conjecture (Conjecture 1.4), we get that (as in the somewhere dense case) H is either bipartite or has bounded odd-girth.

In the reminding case (where $\mathcal{C}$ is a bounded expansion class), we prove that first-order definable coloring correspond to restricted dualities, hence are exactly defined by a sentence expressing that no homomorphism exists from one of the connected graphs belonging to a finite set.

This study naturally leads to the following conjecture:

Conjecture 1.3. Let $\mathcal{C}$ be a hereditary addable topologically closed class of graphs. The following properties are equivalent:

1. for every integer p there is a non-bipartite graph H_p of odd-girth strictly greater than $2p + 1$ and a first order definable class $\mathcal{D}_p$ such that a graph $G \in \mathcal{C}$ is H_p-colorable if and only if $G \in \mathcal{D}_p$. Explicitly, there exists a formula Φ_p such that for every graph $G \in \mathcal{C}$ holds

$$(G \vDash \Phi_p) \qquad \Longleftrightarrow \qquad (G \to H_p);$$

2. the class $\mathcal{C}$ has bounded expansion.

In this paper, we make a significant progress toward a solution to Conjecture 1.3. We state a structural conjecture, Conjecture 1.4, about nowhere dense classes that fail to be bounded expansion classes, which expresses that such classes are characterized by the existence (as subgraphs of graphs in the class) of p-subdivisions of graphs with arbitrarily large chromatic number and girth. More precisely:

Conjecture 1.4. Let C be a monotone nowhere dense class that does not have bounded expansion. Then there exists an integer p such that C includes p-subdivisions of graphs with arbitrarily large chromatic number and girth.

Our main result toward a characterization of first-order definable colorings is the following reduction.

Theorem 1.5. *Let C be a hereditary topologically closed class of graphs that is somewhere dense or has bounded expansion. Then the following properties are equivalent:*

1. *for every integer p there is a non-bipartite graph H_p of odd-girth strictly greater than $2p + 1$ such that H_p-coloring is first-order definable on C;*
2. *the class C has bounded expansion.*

Moreover, if Conjecture 1.4 holds then the statement holds for every hereditary addable topologically closed class of graphs, that is: if Conjecture 1.4 holds then Conjecture 1.3 also holds.

In support to Conjecture 1.4, we prove (Proposition 2.3) that it would follow from a positive solution to any of the following two well known conjectures.

Conjecture 1.6 (Erdős and Hajnal [13]). For every integers g and n there exists an integer $N = f(g, n)$ such that every graph G with chromatic number at least N has a subgraph H with girth at least g and chromatic number at least n.

The case $g = 4$ of the conjecture was proved by Rödl [38], while the general case is still open. Remark that the existence of graphs of both arbitrarily high chromatic number and high girth is a well known result of Erdős [12].

Conjecture 1.7 (Thomassen [42]). For all integers c, g there exists an integer $f(c, g)$ such that every graph G of average degree at least $f(c, g)$ contains a subgraph of average degree at least c and girth at least g.

The case $g = 4$ of this conjecture is a direct consequence of the simple fact that every graph can be made bipartite by deleting at most half of its edges. The case $g = 6$ has been proved in [24].

Although we do not settle Conjecture 1.3, our study led us to the two following characterization theorems of classes that have all restricted dualities.

Theorem 1.8. *Let C be a topologically closed class of graphs. The following properties are equivalent:*

1. *the class C has bounded expansion;*
2. *the class C has all restricted dualities;*
3. *for every odd integer g there exists a graph H_g with odd-girth greater than g such that every graph $G \in C$ with odd-girth greater than g has a homomorphism to H_g.*

These results motivate more general studies of classes of relational structures having all restricted dualities, which includes classes of structures whose Gaifman graphs form a class with bounded expansion. In a very general setting, we obtain the following characterization:

Theorem 1.9. *Let C be a class of σ-structures. Then C is bounded and has all restricted dualities if and only if for every integer t there is an integer $N(t)$ such that for every $\mathbf{A} \in C$ there exists a σ-structure $\mathbf{A}_t$ (called t-approximation of $\mathbf{A}$) such that*

- $\mathbf{A}_t$ *has order $|A_t|$ at most $N(t)$,*
- $\mathbf{A} \to \mathbf{A}_t$,
- *every substructure $\mathbf{F}$ of $\mathbf{A}_t$ with order $|F| < t$ has a homomorphism to $\mathbf{A}$.*

This paper is organized as follows:

In Section 2, we recall the notions needed in the development of our study, in particular class taxonomy and basics on relational structures.

In Section 3, we discuss classes satisfying a homomorphism preservation theorem (HPT). On these classes, first-order definable H-colorings correspond to finite restricted dualities. Moreover, when the considered class is addable, they correspond to finite restricted dualities with connected templates.

In Section 5, we discuss classes having all restricted dualities, that is classes such that to every finite set $\mathcal{F}$ of connected templates correspond a graph H such that a graph G in the class is H-colorable if and only if none of the templates in $\mathcal{F}$ has a homomorphism to G.

2   Taxonomy of Classes of Graphs

In the following, we denote by $\mathcal{G}raph$ the class of all finite graphs. A class of graphs C is *monotone* (respectively, *hereditary*, *topologically closed*) if every subgraph (respectively, every induced subgraph, every subdivision) of a graph in C also belongs to C. Notice that if a class C is both hereditary and topologically closed it is also monotone: If H is a subgraph of a graph $G \in \mathcal{C}$, the graph H is an induced subgraph of the graph $G' \in \mathcal{C}$ obtained from G by subdividing every edge not in H, hence $H \in \mathcal{C}$. For a graph G, we denote by $\omega(G)$ its clique number, by $\chi(G)$ its chromatic number, and by $\overline{d}(G)$ the average degree of its vertices. By extension, for a class of graphs $\mathcal{C}$ we define

$$\omega(\mathcal{C}) = \sup\{\omega(G),\ G \in \mathcal{C}\}$$
$$\chi(\mathcal{C}) = \sup\{\chi(G),\ G \in \mathcal{C}\}$$
$$\overline{d}(\mathcal{C}) = \sup\{\overline{d}(G),\ G \in \mathcal{C}\}$$

We proposed in [29, 31, 34] a general classification scheme for graph classes which is based on the density of shallow (topological) minors (we refer the interested reader to the monography [35]). This classification can be defined in several very different ways and we give here one of the simplest definitions, which relates to subdivision:

A *subdivision* (respectively, a *k-subdivision*) of an edge $e = \{u, v\}$ of a graph G consists in replacing the edge e by a path (respectively, by a path of length $k + 1$) with endpoints u and v. A *subdivision* of a graph G is a graph H resulting from G by subdividing edges; the graph H is the *k-subdivision* of G if it has been obtained from G by k-subdividing all the edges (*i.e.* if all the edges of G have been replaced by paths of length $k + 1$). The graph H is a $\leq k$-*subdivision* of H if it has been obtained by replacing each edge of G by a path of length at most $k + 1$.

Let p be a half-integer. A graph H is a *shallow topological minor* of a graph G *at depth* p if some $\leq 2p$-subdivision of H is a subgraph of G; the set of all shallow topological minors of G at depth p is denoted by $G \widetilde{\triangledown} p$ and, more generally, $\mathcal{C} \widetilde{\triangledown} p$ denotes the class of all shallow topological minors at depth p of graphs in $\mathcal{C}$.

A class of undirected graphs $\mathcal{C}$ is *somewhere dense* if there exists an integer p such that the p-th subdivision of every finite graph H may be found as a subgraph of some graph in $\mathcal{C}$; it is *nowhere dense* otherwise.

In other words, the class $\mathcal{C}$ is nowhere dense if

$$\forall p \in \mathbb{N}, \qquad \omega(\mathcal{C} \widetilde{\triangledown} p) < \infty.$$

A particular type of nowhere dense classes will be of particular importance in this paper: A class $\mathcal{C}$ has *bounded expansion* [29] if

$$\forall p \in \mathbb{N}, \qquad \overline{d}(\mathcal{C} \triangledown p) < \infty.$$

Among the numerous equivalent characterizations that can be given for the property of having bounding expansion, we will make use of a characterization based on the chromatic numbers of the shallow topological minors of the graphs in the class. This characterization can be deduced from the following result of Dvořák [9,10] (see also [35]):

Lemma 2.1. *Let $c \geq 4$ be an integer and let G be a graph with minimum degree $d > 56(c-1)^2 \frac{\log(c-1)}{\log c - \log(c-1)}$. Then the graph G contains a subgraph G' that is the 1-subdivision of a graph with chromatic number c.*

Hence the following characterization of classes having bounded expansion:

Theorem 2.2. *A class C has bounded expansion if and only if it holds*

$$\forall p \in \mathbb{N}, \qquad \chi(C \, \widetilde{\triangledown} \, p) < \infty. \tag{2.1}$$

Proof. According to Lemma 2.1, for every graph G and every integer p there exists an integer C such that:

$$\overline{d}(G \, \widetilde{\triangledown} \, p) \leq C \, \chi(G \, \widetilde{\triangledown} \, (2p + 1/2))^4.$$

Moreover, as every graph G is $(\lfloor \overline{d}(G \, \widetilde{\triangledown} \, 0) \rfloor + 1)$-colorable, every graph in $G \, \widetilde{\triangledown} \, p$ is $(\lfloor \overline{d}(G \, \widetilde{\triangledown} \, p) \rfloor + 1)$-colorable, that is:

$$\overline{d}(G \, \widetilde{\triangledown} \, p) \geq \chi(G \, \widetilde{\triangledown} \, p) - 1.$$

The result follows from these two inequalities. $\square$

Thus we see that parameters $\overline{d}$ and χ can be used to define bounded expansion classes, although nowhere dense classes are defined by means of the parameter ω.

Characterizing nowhere dense classes that do not have bounded expansion in a structural way is challenging, and thus we proposed Conjecture 1.4, from which Conjecture 1.3 would follow. We now prove that Conjecture 1.4 follows from any of the Conjecture 1.6 (by Erdős and Hajnal) or Conjecture 1.7 (by Thomassen):

Proposition 2.3. *If either Conjecture 1.6 or Conjecture 1.7 holds, then so does Conjecture 1.4.*

Proof. Let C be a nowhere dense class of graphs which is not a bounded expansion class. By definition of a bounded expansion class, there exists an integer q such that C includes $\leq q$-subdivisions of graphs with arbitrarily large average degree hence (by standard pigeon-hole argument) there is an integer p such that C includes exact q-subdivisions of graphs with arbitrarily large average degree.

Assume Conjecture 1.6 holds. Define $p = 2q + 1$, and let $g, n \in \mathbb{N}$. According to the statement of the conjecture, there exists N such that every graph with chromatic number at least N has a subgraph with girth at least g and chromatic number at least n. We can assume $N \geq 4$. Let $d \geq 56(N - 1)^2 \frac{\log(N-1)}{\log N - \log(N-1)}$. Let $G \in C$ be such that G includes the q-subdivision of a graph with average degree at least $2d$ hence the q-subdivision of a graph H with minimum degree at least d. According to Lemma 2.1, H has a subgraph H' that is the 1-subdivision of a graph with chromatic number N. It follows that G has a subgraph which is a p-subdivision of a graph K with chromatic number N. According to Conjecture 1.6, the graph K has a subgraph K' which has chromatic number at least n and girth at least g. It follows that G contains the p-subdivision of a graph with chromatic number at least n and girth at least g. Thus Conjecture 1.4 holds.

Assume that Conjecture 1.7 holds. Define $p = 2q + 1$, and let $g, n \in \mathbb{N}$. Let $d \geq 56(n - 1)^2 \frac{\log(n-1)}{\log n - \log(n-1)}$. According to the statement of the conjecture, there exists N such that every graph with average degree at least N has a subgraph with girth at least $2g + 1$ and average degree at least $2d$ hence a subgraph with girth at least $2g + 1$ and minimum degree at least d. Let $G \in C$ be such that G includes the q-subdivision of a graph with average degree at least N. Then G has a subgraph which is the q-subdivision of a graph H with minimum degree at least d and girth at least $2g + 1$. According to Lemma 2.1, H has a subgraph which is the 1-subdivision of a graph with chromatic number at least n. This subgraph is a p-subdivision of a graph with girth at least g and chromatic number at least n. Thus Conjecture 1.4 holds. $\qquad\square$

3 Homomorphism Preservation Theorems

Suppose that an H-coloring problem is first-order definable. By this we mean that there is a first-order sentence Φ such that

$$G \to H \qquad \Longleftrightarrow \qquad G \models \Phi.$$

It immediately follows that $\neg\Phi$ is preserved by homomorphisms:

$$G \models \neg\Phi \quad \text{and} \quad G \to G' \qquad \Longrightarrow \qquad G' \models \neg\Phi$$

(for otherwise $G \to G' \to H$ hence $G \models \Phi$, a contradiction).

Such a property suggests that such a formula Φ could be equivalent to a formula with a specific syntactic form. Indeed the classical *Homomorphism Preservation Theorem* (HPT) asserts that a first-order formula is preserved under homomorphisms on all structures if, and only if, it is logically equivalent to an existential-positive formula. The terms "all structures", which means finite and infinite structures, is crucial in the statement of these theorems.

3.1 Finite Structures

It was not known until recently whether HPT would hold when relativized to the finite. In fact other well known theorems relating preservation under some specified algebraic operation and certain syntactic forms, like Łoś-Tarski theorem or Lyndon's theorem, fail in the finite.

However, the finite relativization of the homomorphism preservation has been proved to hold by B. Rossman [39] for general relational structures.

Theorem 3.1 ([39]). *Let ϕ be a first order formula. Then,*

$$\mathbf{G} \to \mathbf{H} \ and \ \mathbf{G} \vDash \phi \quad \Longrightarrow \quad \mathbf{H} \vDash \phi$$

holds for all finite relational structures $\mathbf{G}$ and $\mathbf{H}$ if and only if for finite relational structures ϕ is equivalent to an existential first-order formula.

It follows that for finite relational structures, the only $\mathbf{H}$-coloring problems which are expressible in first-order logic are those for which there exists a finite family $\mathcal{F}$ of finite structures with the property that for every structure $\mathbf{G}$ the following finite homomorphism duality holds:

$$\exists \mathbf{F} \in \mathcal{F} \quad \mathbf{F} \to \mathbf{G} \qquad \Longleftrightarrow \qquad \mathbf{G} \nrightarrow \mathbf{H}. \tag{3.1}$$

In this paper, we will be mostly interested by graphs, although relational structures will be considered in Section 5. Definitions and constructions concerning relational structures are particularly discussed in Section 5.1.

3.2 Nowhere dense classes

If we want to relativize Theorem 3.1, we should consider each relativization as a new problem. The Łoś-Tarski theorem, for instance, holds in general, yet fails when relativized to the finite, but holds when relativized to hereditary classes of structures with bounded degree which are closed under disjoint union [3]. These examples stress again that some properties of structures (in general) and graphs (in particular) need, at times, to be studied in the context of a fixed class, in order to state a relativized version of a general statement which could fail in general.

In this context Atserias, Dawar and Kolaitis defined classes of graphs called *wide*, *almost wide* and *quasi-wide* (*cf.* [6] for instance). It has been proved in [3] that the extension preservation theorem holds in any class

C that is wide, hereditary (*i.e.* closed under taking substructures) and closed under disjoint unions, for instance hereditary classes with bounded degree that are closed under disjoint unions. Also, it has been proved in [4] [5] that the homomorphism preservation theorem holds in any class C that is almost wide, hereditary and closed under disjoint unions. Almost wide classes of graphs include classes of graphs which exclude a minor [23]. In [8] Dawar proved that the homomorphism preservation theorem holds in any hereditary quasi-wide class that is closed under disjoint unions.

Theorem 3.2 ([8]). *Let C be a hereditary addable quasi-wide class of graphs. Then the homomorphism preservation theorem holds for C.*

Moreover, we have proved that hereditary quasi-wide classes of graphs are exactly hereditary nowhere dense classes [31]:

Theorem 3.3. *A hereditary class of graphs C is quasi-wide if and only if it is nowhere dense.*

Thus it follows from Theorems 3.2 and 3.3 that the relativization of the homomorphism preservation theorem holds for every hereditary addable nowhere dense class of graphs. But nowhere dense classes are not the only classes with relativized homomorphism preservation theorem. In the next section we show HPT also holds for some nowhere dense classes.

3.3 Somewhere dense classes

We now show that relativized homomorphism preservation theorems are preserved by particular interpretations, from which will deduce that relativized homomorphism preservation theorems hold for the classes $\text{Sub}_q(\mathcal{G}raph)$ of all q-subdivisions of (simple) finite graphs. This is of particular interest as somewhere dense classes (*i.e.* classes which fail to be nowhere dense) are characterized by containment of classes $\text{Sub}_q(\mathcal{G}raph)$ for some q.

In the framework of the model theoretical notion of *interpretation* (see, for instance [26, pp. 178-180]), we can construct the q-subdivision $\mathsf{I}(G)$ of a graph G by means of first-order formulas on the q-tuples of vertices of G:

- vertices of $\mathsf{I}(G)$ are the equivalence classes x of the $(q + 1)$-tuples $(v_1, \ldots, v_{q+1})$ with form

$$(\overbrace{u, \ldots, u}^{j}, \overbrace{v, \ldots, v}^{q+1-j})$$

where u and v are adjacent vertices in G (and $0 \leq j \leq q + 1$), where tuples of the form

$$(\overbrace{u, \ldots, u}^{j}, \overbrace{v, \ldots, v}^{q+1-j}) \text{ and } (\overbrace{v, \ldots, v}^{q+1-j}, \overbrace{u, \ldots, u}^{j})$$

are identified;

- edges of $\mathsf{I}(G)$ are those pairs $\{x, y\}$ where x and y have representative of the form

$$(\overbrace{u, \ldots, u}^{j}, \overbrace{v, \ldots, v}^{q+1-j}) \text{ and } (\overbrace{u, \ldots, u}^{j+1}, \overbrace{v, \ldots, v}^{q-j})$$

(for some $u, v \in G$ and $0 \leq j \leq q$).

The main interest of such a logical construction (called *interpretation*) lies in the following property:

Proposition 3.4 (See, for instance [26], p. 180).   *For every first-order formula $\phi[v_1, \ldots, v_k]$ there exists a formula $\mathsf{I}(\phi)[\overline{w}_1, \ldots, \overline{w}_k]$ with $k(q + 1)$ free variables (each $\overline{w}_i$ represents a succession of $(q + 1)$ free variables) such that for every graph G and every $(x_1, \ldots, x_k) \in \mathsf{I}(G)^k$ the three following conditions are equivalent:*

1. $\mathsf{I}(G) \vDash \phi[x_1, \ldots, x_k]$;
2. *there exist $\overline{b}_1 \in x_1, \ldots, \overline{b}_k \in x_k$ such that $G \vDash \mathsf{I}(\phi)[\overline{b}_1, \ldots, \overline{b}_k]$;*
3. *for all $\overline{b}_1 \in x_1, \ldots, \overline{b}_k \in x_k$ it holds $G \vDash \mathsf{I}(\phi)[\overline{b}_1, \ldots, \overline{b}_k]$.*

In particular, it holds:

Corollary 3.5. *For every sentence (i.e. closed first order formula) ϕ (in the language of graphs) there exists a sentence ψ such that for every graph G we have*

$$G \vDash \psi \quad \Longleftrightarrow \quad \text{Sub}_{2p}(G) \vDash \phi. \tag{3.2}$$

Lemma 3.6. *If the homomorphism preservation theorem holds for a hereditary class of graphs C, it also holds for the class $\text{Sub}_q(C)$ of all q-subdivisions of the graphs in C.*

Proof. If q is odd then the property is obvious as C contains at most two homomorphism equivalence classes, the one of K_1 and the one of K_2. Hence we can assume q is even and we define $p = q/2$.

Let ϕ be a sentence preserved by homomorphisms on $\text{Sub}_{2p}(C)$, where C is a hereditary class of graphs on which the homomorphism preservation theorem holds. Then we shall prove that there exists a finite family of $2p$-subdivided graphs $\mathcal{F}$, all of which satisfy ϕ, and such that for any graph G it holds

$$\mathrm{Sub}_{2p}(G) \vDash \phi \quad \Longleftrightarrow \quad \exists F \in \mathcal{F} \quad \mathrm{Sub}_{2p}(F) \to \mathrm{Sub}_{2p}(G). \quad (3.3)$$

According to Corollary 3.5 there exists a sentence ψ such that for every graph G it holds

$$G \vDash \psi \quad \Longleftrightarrow \quad \mathrm{Sub}_{2p}(G) \vDash \phi.$$

Assume that $G \vDash \psi$ and $G \to H$, with $G, H \in \mathcal{C}$. Then $\mathrm{Sub}_{2p}(G) \vDash \phi$ and $\mathrm{Sub}_{2p}(G) \to \mathrm{Sub}_{2p}(H)$. As ϕ is preserved by homomorphisms on $\mathrm{Sub}_{2p}(\mathcal{C})$ we get $\mathrm{Sub}_{2p}(H) \vDash \phi$ hence $H \vDash \psi$. Thus ψ is preserved by homomorphisms on $\mathcal{C}$. As the homomorphism preservation theorem holds by assumption on $\mathcal{C}$, ψ is equivalent on $\mathcal{C}$ with a positive first-order formula, that is: there exits a finite family $\mathcal{F}_0$ of finite graphs such that for every $G \in \mathcal{C}$ it holds:

$$G \vDash \psi \quad \Longleftrightarrow \quad \exists F \in \mathcal{F}_0 \quad F \to G.$$

Moreover, by considering the subgraphs induced by the homomorphic images of the graphs $F \in \mathcal{F}_0$ and as $\mathcal{C}$ is hereditary, we can assume $\mathcal{F}_0 \subseteq \mathcal{C}$. Thus every $F \in \mathcal{F}_0$ satisfies ψ hence the $2p$-subdivision of the graphs in $\mathcal{F}_0$ satisfy ϕ. Let $\mathcal{F}$ be the set of the $2p$-subdivisions of the graphs in $\mathcal{F}_0$. As ϕ is preserved by homomorphisms on $\mathrm{Sub}_{2p}(\mathcal{C})$ it follows that for every graph $G \in \mathcal{C}$ if there exists $F \in \mathcal{F}$ such that $F \to \mathrm{Sub}_{2p}(G)$ then $\mathrm{Sub}_{2p}(G)$ satisfies ϕ. Conversely, if $\mathrm{Sub}_{2p}(G)$ satisfies ϕ for some $G \in \mathcal{C}$ then G satisfies ψ, thus there exists $F \in \mathcal{F}_0$ such that $F \to G$ hence $\mathrm{Sub}_{2p}(F) \to \mathrm{Sub}_{2p}(G)$. $\qquad\square$

We deduce this extension of Rossman's theorem to the class of p-subdivided graphs:

Corollary 3.7. *For every integer p, the homomorphism preservation theorem holds for* $\mathrm{Sub}_p(\mathcal{G}raph)$.

For a discussion on relativization of the homomorphism preservation theorem, we refer the reader to [35, Chapter 10].

We summarize below the results obtained in this section:

Lemma 3.8. *Let $\mathcal{C}$ be a hereditary class of graphs. Assume H-coloring is first-order definable on $\mathcal{C}$.*

- *If C is topologically closed and somewhere dense, then there exist an integer p (independent of H) and a finite set $\mathcal{F}$ of finite graphs such that for every graph G it holds*

$$\mathrm{Sub}_{2p}(G) \nrightarrow H \iff \exists F \in \mathcal{F} : F \to \mathrm{Sub}_{2p}(G). \quad (3.4)$$

- *If C is addable and nowhere dense, then there exists a finite set $\mathcal{F}$ of finite graphs such that for every graph $G \in C$ it holds*

$$G \nrightarrow H \iff \exists F \in \mathcal{F} : F \to G. \quad (3.5)$$

Proof. If C is topologically closed and somewhere dense, then there exist an integer p such that the class $\mathcal{S}_{2p}$ of all $2p$-subdivisions of finite graphs is a subclass of C. As H-coloring is first-order definable on C (thus on $\mathcal{S}_{2p}$) and as the homomorphism preservation theorem holds for $\mathcal{S}_{2p}$ (according to Corollary 3.7), H-coloring may be expressed on $\mathcal{S}_{2p}$ by an existential first-order formulas, that is there exists a finite set $\mathcal{F}$ of finite graphs such that for every graph G equivalence (3.4) holds.

If C is addable and nowhere dense, then the existence of a finite set $\mathcal{F}$ of finite graphs such that for every graph $G \in C$ equivalence (3.5) holds immediately follows from Theorem 3.2. □

4 Connectivity of Forbidden Graphs

Homomorphism preservation theorems allow to reduce the study of first-order colorings of a class C to the study of finite restricted dualities of C, that is of pairs $(\mathcal{F}, H)$, where $\mathcal{F}$ is a finite set of finite graphs, where H is a graph, and where it holds

$$\forall G \in C : \quad (\exists F \in \mathcal{F} : F \nrightarrow G) \iff (G \to H). \quad (4.1)$$

In general, it is not required that $F \nrightarrow H$ when $F \in \mathcal{F}$, nor is it required that the graphs in $\mathcal{F}$ are connected. We shall see that if the class C is addable or monotone then we can require the graphs in $\mathcal{F}$ to be connected, and that if C is monotone we can further require every $F \in \mathcal{F}$ belongs to C (hence cannot be homomorphic to H).

For a graph G, we define $\mathrm{Pre}(G)$ has the set of all the graphs G' obtained from G by identifying two vertices of G. Note that we have the following property:

Lemma 4.1. *Let F, G be graphs. Then $F \to G$ if and only if either F is isomorphic to a subgraph of G, or there is $F' \in \mathrm{Pre}(F)$ such that $F' \to G$.*

Proof. Assume $f : F \to G$ is a homomorphism. Either f is injective and F is isomorphic to a subgraph of G, or at least two vertices of F are identified by f and thus there is $F' \in \mathrm{Pre}(F)$ such that $F' \to G$. Conversely, if F is isomorphic to a subgraph of G or if there is $F' \in \mathrm{Pre}(F)$ such that $F' \to G$ then obviously $F \to G$. □

Let $\mathcal{C}$ be a class of graphs and let $(\mathcal{F}, H)$ be a restricted duality of $\mathcal{C}$. We say that the set $\mathcal{F}$ is *minimal* if

- every element of $\mathcal{F}$ is a *core* (that is a graph F such that every homo-morphism $F \to F$ is an automorphism);
- for any proper subset of $\mathcal{F}'$ of $\mathcal{F}$, the pair $(\mathcal{F}', H)$ is not a restricted duality of $\mathcal{C}$,
- and for any $F \in \mathcal{F}$, the pair $(\mathcal{F} \setminus \{F\} \cup \mathrm{Pre}(F), H)$ is not a restricted duality of $\mathcal{C}$.

It is clear that we can restrict our attention to minimal restricted dualities, as if $(\mathcal{F}, H)$ is a restricted duality of a class $\mathcal{C}$ then there exists minimal $\mathcal{F}'$ such that $(\mathcal{F}', H)$ is a restricted duality of $\mathcal{C}$.

Lemma 4.2. *Let $(\mathcal{F}, H)$ be a restricted duality of $\mathcal{C}$, with $\mathcal{F}$ minimal.*

- *If the class $\mathcal{C}$ is addable, then every graph in $\mathcal{F}$ is connected.*
- *If the class $\mathcal{C}$ is monotone, then every graph in $\mathcal{F}$ is connected and $\mathcal{F} \subseteq \mathcal{C}$ (hence $F \nrightarrow H$ holds for every $F \in \mathcal{F}$).*

Proof. Assume $\mathcal{C}$ is addable, and assume for contradiction that $F_1 + F_2 \in \mathcal{F}$. By minimality of $\mathcal{F}$, neither $(\mathcal{F} \setminus \{F_1 + F_2\} \cup \{F_1\}, H)$ nor $(\mathcal{F} \setminus \{F_1 + F_2\} \cup \{F_2\}, H)$ are restricted dualities of $\mathcal{C}$. Hence there exist $G_1, G_2 \in \mathcal{C}$ such that $F_1 \nrightarrow G_1, F_2 \to G_1$ (hence $G_1 \to H$), $F_1 \to G_2, F_2 \nrightarrow G_2$ (hence $G_2 \to H$). As $\mathcal{C}$ is addable, $G_1 + G_2 \in \mathcal{C}$. But $F_1 + F_2 \to G_1 + G_2$, what contradicts $G_1 + G_2 \to H$.

Assume $\mathcal{C}$ is monotone, and assume $F \in \mathcal{F}$. By minimality of $\mathcal{F}$, there exists $G \in \mathcal{C}$ such that $F \to G$ but no graph $F' \in \mathrm{Pre}(F)$ is homomorphic to G. Thus, according to Lemma 4.1, F is isomorphic to a subgraph of G hence, as $\mathcal{C}$ is monotone, $F \in \mathcal{C}$. Assume for contradiction that $F = F_1 + F_2$. Then $F_1, F_2 \in \mathcal{C}$ and none of F_1, F_2 is homomorphic to the other. By minimality of $\mathcal{F}$ it follows that $F_1 \to H, F_2 \to H$ hence $F_1 + F_2 \to H$, contradicting $F_1 + F_2 \in \mathcal{F}$. □

5 Restricted Dualities

As restricted dualities appear as the central notion when dealing with first-order definable coloring, we take time to define and characterize classes with all restricted dualities in the more general framework of re-lational structures.

5.1   Classes of Relational Structures

We recall some basic definitions, notations and result of model theory. Our terminology is standard, *cf.* [11,26]:

A *signature* σ is a finite set of relation symbols, each with a specified arity. A σ-*structure* **A** consists of a *universe* A, or *domain*, and an *interpretation* which associates to each relation symbol $R \in \sigma$ of some arity r, a relation $R^{\mathbf{A}} \subseteq A^r$.

A σ-structure **B** is a *substructure* of **A** if $B \subseteq A$ and $R^{\mathbf{B}} \subseteq R^{\mathbf{A}}$ for every $R \in \sigma$. It is an *induced substructure* if $R^{\mathbf{B}} = R^{\mathbf{A}} \cap B^r$ for every $R \in \sigma$ of arity r. Notice the analogy with the graph-theoretical concept of subgraph and induced subgraph. A substructure **B** of **A** is *proper* if $\mathbf{A} \neq \mathbf{B}$. If **A** is an induced substructure of **B**, we say that **B** is an *extension* of **A**. If **A** is a proper induced substructure, then **B** is a *proper extension*. If **B** is the disjoint union of **A** with another σ-structure, we say that **B** is a *disjoint extension* of **A**. If $S \subseteq A$ is a subset of the universe of **A**, then $\mathbf{A} \cap S$ denotes the *induced substructure generated by* S; in other words, the universe of $\mathbf{A} \cap S$ is S, and the interpretation in $\mathbf{A} \cap S$ of the r-ary relation symbol R is $R^{\mathbf{A}} \cap S^r$.

The *Gaifman graph* Gaifman(**A**) of a σ-structure **A** is the graph with vertex set A in which two vertices $x \neq y$ are adjacent if and only if there exists a relation R of arity $k \geq 2$ in σ and $v_1, \ldots, v_k \in A$ such that $\{x, y\} \subseteq \{v_1, \ldots, v_k\}$ and $(v_1, \ldots, v_k) \in R^{\mathbf{A}}$.

A *block* of a σ-structure **A** is a tuple $(R, x_1, \ldots, x_k)$ such that $R \in \sigma$ has arity k and $(x_1, \ldots, x_k) \in R^{\mathbf{A}}$. The *incidence graph* Inc(**A**) is the bipartite graph (A, B, E) where A is the universe of **A**, B is the set of all *blocks* of **A**, and E is the set of the pairs $\{(R, x_1, \ldots, x_k), y\} \subseteq B \times A$ such that $y \in \{x_1, \ldots, x_k\}$. Thus for us Inc(**A**) is a simple graph. No multiple edges are needed for our purposes.

A *homomorphism* $\mathbf{A} \to \mathbf{B}$ between two σ-structure is defined as a mapping $f : A \to B$ which satisfies for every relational symbol $R \in \sigma$ the following:

$$(x_1, \ldots, x_k) \in R^{\mathbf{A}} \quad \Longrightarrow \quad (f(x_1), \ldots, f(x_k)) \in R^{\mathbf{B}}.$$

The class of all σ-structures is denoted by Rel(σ).

The definition of bounded expansion extends to classes of relational structures: a class $\mathcal{C}$ of relational structures has *bounded expansion* if the class of the Gaifman graphs of the structures in $\mathcal{C}$ has bounded expansion. It is immediate that two relational structures have the same Gaifman graph if they have the same incidence graph, but that the converse does not hold in general. For a class of relational structures $\mathcal{C}$, denote by Inc($\mathcal{C}$) the class of all the incidence graphs Inc($\mathbf{A}$) of the relational structures $\mathbf{A} \in \mathcal{C}$.

Proposition 5.1 ([35]). *Assume that the arities of the relational symbols in σ are bounded, and let $\mathcal{C}$ be an infinite class of σ-structures. Then the class $\mathcal{C}$ has bounded expansion if and only if the class* Inc($\mathcal{C}$) *has bounded expansion.*

5.2 Classes with all restricted dualities

A class of σ-structures $\mathbf{A}$ has *all restricted dualities* if every non-empty connected σ-structure has a restricted dual for $\mathcal{C}$, that is: for every non-empty connected σ-structure $\mathbf{F}$ there exists a σ-structure $\mathbf{D}$ such that $\mathbf{F} \nrightarrow \mathbf{D}$ and
$$\forall \mathbf{A} \in \mathcal{C} : \qquad (\mathbf{F} \to \mathbf{A}) \qquad \Longleftrightarrow \qquad (\mathbf{A} \nrightarrow \mathbf{D}).$$

Note that this definition implies that also for any finite set $\mathbf{F}_1, \mathbf{F}_2, \ldots, \mathbf{F}_t$ of connected σ-structures there exists a σ-structure $\mathbf{D}$ such that $\mathbf{F}_i \nrightarrow \mathbf{D}$ (for $1 \leq i \leq t$) and
$$\forall \mathbf{A} \in \mathcal{C} : \qquad (\exists i \leq t : \mathbf{F}_i \to \mathbf{A}) \qquad \Longleftrightarrow \qquad (\mathbf{A} \nrightarrow \mathbf{D}).$$

For a structure $\mathbf{A}$ and an integer t, define $\Theta^t(\mathbf{A})$ as the minimum order of a structure $\mathbf{B}$ such that

- $\mathbf{A} \to \mathbf{B}$,
- every substructure $\mathbf{F}$ of $\mathbf{B}$ with order $|F| < t$ has a homomorphism to $\mathbf{A}$.

Intuitively, such a structure $\mathbf{B}$ can be seen as approximate core of $\mathbf{A}$: For $t \geq |B|$, $\mathbf{A}$ and $\mathbf{B}$ are homomorphism-equivalent and $\mathbf{B}$ is the core of $\mathbf{A}$ (alternately, $\mathbf{B}$ is the minimal retract of $\mathbf{A}$). A structure $\mathbf{B}$ with the above properties and order $\Theta^t(\mathbf{A})$ is called a *t-approximation* of (the homomorphism equivalence class of) $\mathbf{A}$.

It appears that existence of a uniform approximation is equivalent for a class to having all restricted dualities. This is formalized by Theorem 1.9, stated in the introduction. Theorem 1.9 will be proved now:

Proof of Theorem 1.9. Assume $\mathcal{C}$ is bounded and has all restricted dualities and let $t \in \mathbb{N}$ be an integer. Let $\mathbf{Z}$ be a strict bound of $\mathcal{C}$, that is a structure such that for every $\mathbf{A} \in \mathcal{C}$ it holds $\mathbf{A} \to \mathbf{Z}$ but $\mathbf{Z} \nrightarrow \mathbf{A}$. As the sequence $(\Theta^t(\mathbf{A}))_{t \in \mathbb{N}}$ is obviously non-decreasing, we may assume without loss of generality that $t \geq |Z|$. For a structure $\mathbf{A} \in \mathcal{C}$, let $\mathcal{F}_t(\mathbf{A})$ be the set of all connected cores $\mathbf{T}$ of order at most t such that $\mathbf{T} \nrightarrow \mathbf{A}$. This set is not empty as it contains the core of $\mathbf{Z}$. For $\mathbf{T} \in \mathcal{F}_t(\mathbf{A})$, let

$\mathbf{D_T}$ be the dual of $\mathbf{T}$ relative to C and let $\mathbf{A}'$ be the product of all the $\mathbf{D_T}$ for $\mathbf{T} \in \mathcal{F}_t(\mathbf{A})$. First notice that for every $\mathbf{T} \in \mathcal{F}_t(\mathbf{A})$ we have $\mathbf{T} \nrightarrow \mathbf{A}$ hence $\mathbf{A} \rightarrow \mathbf{D_T}$. It follows that $\mathbf{A} \rightarrow \mathbf{A}'$. Let $\mathbf{T}'$ be a connected substructure of order at most t of $\mathbf{A}'$. Suppose for a contradiction that $\mathbf{T}' \nrightarrow \mathbf{A}$. Then $\mathrm{Core}(\mathbf{T}') \in \mathcal{F}_t(\mathbf{A})$ hence $\mathbf{A}' \rightarrow \mathbf{D_{T'}}$ thus $\mathbf{T}' \nrightarrow \mathbf{A}'$ (as for otherwise $\mathbf{T}' \rightarrow \mathbf{D_{T'}}$), a contradiction. Thus $\mathbf{T}' \rightarrow \mathbf{A}$. It follows that $\Theta^t(\mathbf{A}) \leq |\mathbf{A}'| \leq C(t)$ for some suitable finite constant $C(t)$ independent of $\mathbf{A}$ (for instance, one can choose $C(t)$ to be the product of the orders of all the duals relative to C of connected cores of order at most t).

Conversely, assume that we have $\sup_{\mathbf{A} \in C} \Theta^t(\mathbf{A}) < \infty$ for every $t \in \mathbb{N}$. The class C is obviously bounded by the disjoint union of all non-isomorphic minimal order 1-approximations of the structures in C. Let $\mathbf{F}$ be a connected σ-structure, let $t \geq |\mathbf{F}|$, and let $\mathcal{D}$ be a set of t-approximations of all the structures $\mathbf{A} \in C$ such that $\mathbf{F} \nrightarrow \mathbf{A}$. As all the $\Theta^t(\mathbf{A})$ are bounded by some constant $C(t)$, the set $\mathcal{D}$ is finite. If $\mathcal{D}$ is empty, let $D_t(\mathbf{F})$ be the empty substructure. Otherwise, let $D_t(\mathbf{F})$ be the disjoint union of all the graphs in $\mathcal{D}$. First notice that $\mathbf{F} \nrightarrow D_t(\mathbf{F})$ as for otherwise $\mathbf{F}$ would have a homomorphism to some structure in $\mathcal{D}$ (as $\mathbf{F}$ is connected), that is to some t-approximation $\mathbf{B}'$ of a structure $\mathbf{B}$ such that $\mathbf{F} \nrightarrow \mathbf{B}$ (this would contradict $\mathbf{F} \rightarrow \mathbf{B}'$). Also, if $\mathbf{F} \rightarrow \mathbf{A}$ then $\mathbf{A} \nrightarrow D_t(\mathbf{F})$ (for otherwise $\mathbf{F} \rightarrow D_t(\mathbf{F})$) and if $\mathbf{F} \nrightarrow \mathbf{A}$ then $\mathcal{D}$ contains a t-approximation $\mathbf{A}'$ of $\mathbf{A}$ thus $\mathbf{A} \rightarrow D_t(\mathbf{F})$. Altogether, $D_t(\mathbf{F})$ is a dual of $\mathbf{F}$ relative to C. □

We proved in [30] that bounded expansion classes have all restricted dualities:

Theorem 5.2. *Let C be a class with bounded expansion. Then for every connected graph F there exists a graph D such that (F, D) is a restricted homomorphism duality for C:*

$$\forall G \in C \qquad (F \rightarrow G) \quad \Longleftrightarrow \quad (G \nrightarrow H). \qquad (5.1)$$

Theorem 5.2 naturally extends to relational structures by considering Gaifman graphs. We sketch a proof of this generalization, which is based on the above Theorem 1.9.

Theorem 5.3. *Let $\mathcal{K}$ be a class of relational structures. If the class of the Gaifman graphs of the structures in $\mathcal{K}$ has bounded expansion then the class $\mathcal{K}$ has all restricted dualities.*

Sketch of the proof. Let $\mathcal{K}$ be a class of relational structures. Assume the class of the Gaifman graphs of the structures in $\mathcal{K}$ has bounded expansion.

Let $\mathbf{A} \in \mathcal{K}$, and let $t \in \mathbb{N}$ be at least as large as the maximum arity of a relation in the signature of $\mathbf{A}$.

The *tree-depth* $\mathrm{td}(G)$ of a graph G is the minimum height of a rooted forest whose closure includes G as a subgraph. One of the most interesting properties of tree-depth is that there exists a function $F : \mathbb{N} \to \mathbb{N}$ with the property that if the Gaifman graph of a structure $\mathbf{B}$ has tree-depth at most t then there exists a homomorphism $f : \mathbf{B} \to \mathbf{B}$ such that $|f(\mathbf{B})| \leq F(t)$ [28]. For integer t, we defined in [28] the graph invariant χ_t as follows: for a graph G, $\chi_t(G)$ is the minimum number of colors needed in a coloring of G such that the union of every subset of $k \leq t$ color classes induces a subgraph with tree-depth at most k (such colorings are called *low tree-depth colorings*). It has been proved in [29] that a class of graphs $\mathcal{C}$ has bounded expansion if and only if for every integer t it holds $\sup\{\chi_t(G) : G \in \mathcal{C}\} < \infty$ (this is related to Theorem 2.2 above).

Consider a coloring c of the Gaifman graph of $\mathbf{A}$ by $N = \chi_t(\mathrm{Gaifman}(\mathbf{A}))$ colors, which is such that the union of every subset of $k \leq t$ color classes induces a subgraph with tree-depth at most k. It follows that for each $I \in \binom{[N]}{t}$ there exists a homomorphism $f_I : \mathbf{A}_I \to \mathbf{A}_I$ such that $|f_I(\mathbf{A}_I)| \leq F(t)$, where $\mathbf{A}_I$ denotes the substructure of $\mathbf{A}$ induced by elements with color in I. Define the equivalence relation $\sim$ on the domain of $\mathbf{A}$ by

$$x \sim y \quad \Longleftrightarrow \quad c(x) = c(y) \text{ and } \forall I \in \binom{[N]}{t} \; f_I(x) = f_I(y).$$

Define the structure $\hat{\mathbf{A}}$ (with same signature as $\mathbf{A}$) whose domain is the set of the equivalence classes $[x] \in A/\sim$, and relations are defined by

$$([x_1], \ldots, [x_{k_i}]) \in R_i^{\hat{\mathbf{A}}} \quad \Longleftrightarrow \quad \forall I \in \binom{[N]}{t} \; (f_I(x_1), \ldots, f_I(x_{k_i})) \in R_i^{\mathbf{A}}.$$

We also define a N-coloration of $\hat{\mathbf{A}}$ by $\hat{c}([x]) = c(x)$. One checks easily that $\hat{\mathbf{A}}$ and $\hat{c}$ are well defined. By construction, $x \mapsto [x]$ is a homomorphism $\mathbf{A} \to \hat{\mathbf{A}}$. Moreover, for every $I \in \binom{[N]}{t}$ the mapping $[x] \mapsto f_I(x)$ is a homomorphism $\hat{\mathbf{A}}_I \to \mathbf{A}_I$ (where $\hat{\mathbf{A}}_I$ is the substructure of $\hat{\mathbf{A}}$ induced by colors in I). It follows that

$$|\Theta^t(\mathbf{A})| \leq |\hat{A}| \leq F(t)^{N^t} \leq F(t)^{\chi_t(\mathrm{Gaifman}(\mathcal{K}))^t}.$$

According to Theorem 1.9, this implies that the class $\mathcal{K}$ has all restricted dualities.   □

For an alternate proof of this Theorem, we refer the reader to [30,35].

5.3 Topologically closed classes of graphs with all restricted dualities

The special case of topologically closed classes of graphs is of particular interest here, and we have in this case a much simpler characterization of the classes that have all restricted dualities. We are now ready to prove Theorem 1.8.

Proof of Theorem 1.8. The proof follows from the next three implications:

- $(1) \Rightarrow (2)$ is a direct consequence of Theorem 5.2.
- $(2) \Rightarrow (3)$ is straightforward (consider for H_g a dual of C_g for C).
- $(3) \Rightarrow (1)$ is proved by contradiction: assume that (3) holds and that C does not have bounded expansion. According to Theorem 2.2 there exists an integer p such that $C \widetilde{\triangledown} p$ has unbounded chromatic number. As C is topologically closed there exists an odd integer $g \geq p$ and a graph $G_0 \in C$ such that G_0 is the $(g-1)$-subdivision of a graph H_0 with chromatic number $\chi(H_0) > |H_g|$. According to (3), there exists a homomorphism $f : G_0 \to H_g$. As $C_g \nrightarrow H_g$, the ends of a path of length g cannot have the same image by f. It follows that any two adjacent vertices in H_0 correspond to branching vertices of G_0 which are mapped by f to distinct vertices of H_g. It follows that $\chi(H_0) \leq |H_g|$, a contradiction.

$\square$

6 On first-order definable H-colorings

In this section we prove our main characterization result on first-order definable colorings, stated in the introduction as Theorem 1.5.

Proof of Theorem 1.5. Assume that the class C is somewhere dense. As C is topologically closed, there exists an integer p such that $\mathrm{Sub}_{2p}(\mathcal{G}raph) \subseteq C$. Assume for contradiction that there exists a non-bipartite graph H (different from K_1) of odd-girth strictly greater than $2p+1$ and a first-order formula Φ such that for every graph $G \in C$ holds

$$(G \vDash \Phi) \quad \Longleftrightarrow \quad (G \to H).$$

According to Corollary 3.7, as $\neg\Phi$ is preserved by homomorphisms on C (hence on $\mathrm{Sub}_{2p}(\mathcal{G}raph)$) it is equivalent on $\mathrm{Sub}_{2p}(\mathcal{G}raph)$ with an existential first-order formula, that is: there exists a finite family $\mathcal{F}$ such that for every graph G it holds:

$$\forall F \in \mathcal{F} \; F \nrightarrow \mathrm{Sub}_{2p}(G) \quad \Longleftrightarrow \quad \mathrm{Sub}_{2p}(G) \to H.$$

Clearly, the graphs in $\mathcal{F}$ are non-bipartite. Let g be the maximum of girth of graphs in $\mathcal{F}$ and let G be a graph with chromatic number $\chi(G) > |H|$ and odd-girth odd $-$ girth$(G) > g$. Then for every $F \in \mathcal{F}$ we have $F \nrightarrow \text{Sub}_{2p}(G)$ hence $\text{Sub}_{2p}(G) \to H$. However, has the odd-girth of H is strictly greater than $2p + 1$ two branching vertices of $\text{Sub}_{2p}(G)$ corresponding to adjacent vertices of G cannot be mapped to a same vertex. It follows that $|H| \geq \chi(G)$, a contradiction.

To the opposite, if $\mathcal{C}$ has bounded expansion, there exists for every integer p a non-bipartite graph H_p of odd-girth strictly greater than $2p+1$ and a first order formula Φ_p such that for every graph $G \in \mathcal{C}$ holds

$$(G \vDash \Phi_p) \quad \Longleftrightarrow \quad (G \to H_p).$$

Indeed, consider for Φ_p the formula asserting that G contains an odd cycle of length at most $2p + 1$, and for H_p the restricted dual of the cycle C_{2p+1} with respect to $\mathcal{C}$ (whose existence follows from Theorem 5.2.

Now assume that Conjecture 1.4 holds, and that $\mathcal{C}$ is a hereditary topologically closed addable nowhere dense class that is not a bounded expansion class. Then there exists an integer p such that $\mathcal{C}$ includes $2p$-subdivisions of graph with arbitrarily large chromatic number and girth. Assume for contradiction that there is a non-bipartite graph H_p of odd-girth strictly greater than $2p + 1$ such that H_p-coloring is first-order definable on $\mathcal{C}$, and let Φ be a formula such that for every $G \in \mathcal{C}$ it holds $G \to H_p$ if and only if $G \models \Phi$. According to Theorem 3.2, there exists a finite family $\mathcal{F}$ of finite graphs such that $G \models H_p$ if and only if no graph in $\mathcal{F}$ is homomorphic to G. The graphs in $\mathcal{F}$ are non-bipartite ($\mathcal{C}$ contains long odd cycles homomorphic to H_p). Let g be the maximum of the odd-girths of the graphs in $\mathcal{F}$. Let $G \in \mathcal{C}$ be a $2p$-subdivision of a graph H with girth greater than g and chromatic number $\chi(H) > |H_p|$. As the girth of G is greater than g, no graph in $\mathcal{F}$ is homomorphic to G hence there exists a homomorphism $f : G \to H_p$. As $C_{2p+1} \nrightarrow G$ it follows that vertices of G linked by a path of length $2p + 1$ are mapped to distinct vertices of H_p hence f defines a homomorphism $H \to K_{|H_p|}$, contradicting $\chi(H) > |H_p|$. $\qquad\square$

References

[1] A. ATSERIAS, *On digraph coloring problems and treewidth duality*, 20th IEEE Symposium on Logic in Computer Science (LICS), 2005, 106–115.

[2] A. ATSERIAS, *On digraph coloring problems and treewidth duality*, European J. Combin. **29** (2008), no. 4, 796–820.

[3] A. ATSERIAS, A. DAWAR and M. GROHE, *Preservation under extensions on well-behaved finite structures*, 32nd International Colloquium on Automata, Languages and Programming (ICALP) (Springer-Verlag, ed.), Lecture Notes in Computer Science, Vol. 3580, 2005, 1437–1449.

[4] A. ATSERIAS, A. DAWAR and P. G. KOLAITIS, *On preservation under homomorphisms and unions of conjunctive queries*, Proceedings of the twenty-third ACM SIGMOD-SIGACT-SIGART symposium on Principles of database systems, ACM Press, 2004, 319 – 329.

[5] A. ATSERIAS, A. DAWAR and P. G. KOLAITIS, *On preservation under homomorphisms and unions of conjunctive queries*, J. ACM **53** (2006), 208–237.

[6] A. DAWAR, *Finite model theory on tame classes of structures*, Mathematical Foundations of Computer Science 2007 (L. Kučera and A. Kučera, eds.), Lecture Notes in Computer Science, vol. 4708, Springer, 2007, 2–12.

[7] A. DAWAR, *On preservation theorems in finite model theory*, Invited talk at the 6th Panhellenic Logic Symposium - Volos, Greece, July 2007.

[8] A. DAWAR, *Homomorphism preservation on quasi-wide classes*, Journal of Computer and System Sciences **76** (2010), 324–332.

[9] Z. DVOŘÁK, *Asymptotical structure of combinatorial objects*, Ph.D. thesis, Charles University, Faculty of Mathematics and Physics, 2007.

[10] Z. DVOŘÁK *On forbidden subdivision characterizations of graph classes*, European J. Combin. **29** (2008), no. 5, 1321–1332.

[11] H.-D. EBBINGHAUS and J. FLUM, "Finite Model Theory", Springer-Verlag, 1996.

[12] P. ERDŐS, *Graph theory and probability*, Canad. J. Math. **11** (1959), no. 1, 34–38.

[13] P. ERDŐS and A. HAJNAL, *On chromatic number of graphs and set-systems*, Acta Math. Acad. Sci. Hungar. **17** (1966), 61–99.

[14] T. FEDER and M. Y. VARDI, *Monotone monadic SNP and constraint satisfaction*, Proceedings of the 25rd Annual ACM Symposium on Theory of Computing (STOC), 1993, 612–622.

[15] T. FEDER and M. Y. VARDI, *The computational structure of monotone monadic SNP and constraint satisfaction: A study through datalog and group theory.*, SIAM J. Comput. **28** (1999), no. 1, 57–104 (English).

[16] T. GALLAI, *On directed graphs and circuits*, Theory of Graphs (Proc. Colloq. Tihany 1966), Academic Press, 1968, 115–118.

[17] M. HASSE, *Zur algebraischen Begründung der Graphentheorie I*, Math. Nachrichten **28** (1964/5), 275–290.

[18] P. HELL and J. NEŠETŘIL, *On the complexity of H-coloring*, J. Combin. Theory Ser. B (1990), no. 48, 92–110.

[19] P. HELL and J. NEŠETŘIL, "Graphs and Homomorphisms", Oxford Lecture Series in Mathematics and its Applications, Vol. 28, Oxford University Press, 2004.

[20] W. HODGES, "Model Theory", Cambridge University Press, 1993.

[21] N. IMMERMAN, *Relational queries computable in polynomial time (extended abstract)*, Proceedings of the Fourteenth Annual ACM Symposium on Theory of Computing (New York, NY, USA), STOC '82, ACM, 1982, 147–152.

[22] P. KOMÁREK, "Good Characterizations of Graphs", Ph.D. thesis, Charles University, Prague, 1988.

[23] M. KREIDLER and D. SEESE, *Monadic NP and graph minors*, Computer Science Logic, Lecture Notes in Computer Science, Vol. 1584, Springer, 1999, 126–141.

[24] D. KÜHN and D. OSTHUS, *Every graph of sufficiently large average degree contains a C_4-free subgraph of large average degree*, Combinatorica **24** (2004), no. 1, 155–162.

[25] G. KUN and J. NEŠETŘIL, *Forbidden lifts (NP and CSP for combinatorialists)*, European. J. Combin. **29** (2008), no. 4, 930–945.

[26] D. LASCAR, "La théorie des modèles en peu de maux", Cassini, 2009.

[27] L. LIBKIN, "Elements of Finite Model Theory", Springer-Verlag, 2004.

[28] J. NEŠETŘIL and P. OSSONA DE MENDEZ, *Tree depth, subgraph coloring and homomorphism bounds*, European Journal of Combinatorics **27** (2006), no. 6, 1022–1041.

[29] J. NEŠETŘIL and P. OSSONA DE MENDEZ, *Grad and classes with bounded expansion I. decompositions*, European Journal of Combinatorics **29** (2008), no. 3, 760–776.

[30] J. NEŠETŘIL and P. OSSONA DE MENDEZ, *Grad and classes with bounded expansion III. restricted graph homomorphism dualities*, European Journal of Combinatorics **29** (2008), no. 4, 1012–1024.

[31] J. NEŠETŘIL and P. OSSONA DE MENDEZ, *First order properties on nowhere dense structures*, The Journal of Symbolic Logic **75** (2010), no. 3, 868–887.

[32] J. NEŠETŘIL and P. OSSONA DE MENDEZ, *From sparse graphs to nowhere dense structures: Decompositions, independence, dualities and limits*, European Congress of Mathematics, European Mathematical Society, 2010, 135–165.

[33] J. NEŠETŘIL and P. OSSONA DE MENDEZ, *Sparse combinatorial structures: Classification and applications*, Proceedings of the International Congress of Mathematicians 2010 (ICM 2010) (Hyderabad, India) (R. Bhatia and A. Pal, eds.), vol. IV, World Scientific, 2010, 2502–2529.

[34] J. NEŠETŘIL and P. OSSONA DE MENDEZ, *On nowhere dense graphs*, European Journal of Combinatorics **32** (2011), no. 4, 600–617.

[35] J. NEŠETŘIL and P. OSSONA DE MENDEZ, *Sparsity (graphs, structures, and algorithms)*, Algorithms and Combinatorics, Vol. 28, Springer, 2012, 465 pages.

[36] J. NEŠETŘIL and A. PULTR, *On classes of relations and graphs determined by subobjects and factorobjects*, Discrete Math. **22** (1978), 287–300.

[37] J. NEŠETŘIL and C. TARDIF, *Duality theorems for finite structures (characterizing gaps and good characterizations)*, Journal of Combinatorial Theory, Series B **80** (2000), 80–97.

[38] V. RÖDL, *On the chromatic number of subgraphs of a given graph*, Proc. Amer. Math. Soc. **64** (1977), no. 2, 370–371.

[39] B. ROSSMAN, *Homomorphism preservation theorems*, J. ACM **55** (2008), no. 3, 1–53.

[40] B. ROY, *Nombre chromatique et plus longs chemins d'un graphe*, Rev. Francaise Informat. Recherche Operationelle **1** (1967), 129–132.

[41] T. J. SCHAEFER, *The complexity of satisfiability problems*, STOC '78 Proceedings of the tenth annual ACM symposium on Theory of computing, 1978, 216–226.

[42] C. THOMASSEN, *Girth in graphs*, Journal of Combinatorial Theory, Series B **35** (1983), 129–141.

[43] C. THOMASSEN, *Five-coloring graphs on the torus*, Journal of Combinatorial Theory, Series B **62** (1994), no. 1, 11 – 33.

[44] M. Y. VARDI, *The complexity of relational query languages*, Proceedings of the 14th ACM Symposium on Theory of Computing, 1982, pp. 137–146.

[45] L. M. VITAVER, *Determination of minimal colouring of vertices of a graph by means of boolean powers of incidence matrix*, Doklady Akad. Nauk SSSR **147** (1962), 758–759.

Combinatorial applications of the subspace theorem

Ryan Schwartz and József Solymosi

Abstract. The Subspace Theorem is a powerful tool in number theory. It has appeared in various forms and been adapted and improved over time. Its applications include diophantine approximation, results about integral points on algebraic curves and the construction of transcendental numbers. But its usefulness extends beyond the realms of number theory. Other applications of the Subspace Theorem include linear recurrence sequences and finite automata. In fact, these structures are closely related to each other and the construction of transcendental numbers. The Subspace Theorem also has a number of remarkable combinatorial applications. The purpose of this paper is to give a survey of some of these applications including bounds on unit distances, sum-product estimates and a result about the structure of complex lines. The presentation will be from the point of view of a discrete mathematician. We will state a number of variants and a corollary of the Subspace Theorem and give a proof of a simplified special case of the corollary which is still very useful for many problems in discrete mathematics.

1 Introduction

The Subspace Theorem is a powerful tool in number theory. It has appeared in various forms and been adapted and improved over time. Its applications include diophantine approximation, results about integral points on algebraic curves and the construction of transcendental numbers. But its usefulness extends beyond the realms of number theory. Other applications of the Subspace Theorem include linear recurrence sequences and finite automata. In fact, these structures are closely related to each other and the construction of transcendental numbers.

The Subspace Theorem also has a number of remarkable combinatorial applications. The purpose of this paper is to give a survey of some of these applications including sum-product estimates and bounds on unit distances. The presentation will be from the point of view of a discrete

Work by József Solymosi was supported by NSERC, ERC-AdG. 321104, and OTKA NK 104183 grants.

mathematician. We will state a number of variants and a corollary of the Subspace Theorem below but we will not prove any of them as the proofs are beyond the scope of this work. However we will give a proof of a simplified special case of the corollary of the Subspace Theorem which is still very useful for many problems in discrete mathematics.

A number of surveys have been given of the Subspace Theorem highlighting its multitude of applications. Notable surveys include those of Bilu [6], Evertse and Schlickewei [20] and Corvaja and Zannier [11]. These surveys give many proofs of results from number theory and algebraic geometry including those mentioned above.

Wolfgang M. Schmidt was the first to state and prove a variant of the Subspace Theorem in 1972 [29]. His theorem has been extended multiple times and has played a very important role in modern number theory. Before we state the Subspace Theorem we need some definitions. A *linear form* is an expression of the form $L(x) = a_1 x_1 + a_2 x_2 + \cdots + a_n x_n$ where $a_1, \ldots, a_n$ are constants and $x = (x_1, \ldots, x_n)$. A collection of linear forms is *linearly independent* if none of them can be expressed as a linear combination of the others. A complex number is *algebraic* if it is a root of a univariate polynomial with rational coefficients. Given $x = (x_1, \ldots, x_n)$ we define the *maximum* norm of x:

$$\|x\| = \max(|x_1|, \ldots, |x_n|).$$

Theorem 1.1 (Subspace Theorem I). *Suppose we have n linearly independent linear forms $L_1, L_2, \ldots, L_n$ in n variables with algebraic coefficients. Given $\varepsilon > 0$, the non-zero integer points $x = (x_1, x_2, \ldots, x_n)$ satisfying*

$$|L_1(x) L_2(x) \ldots L_n(x)| < \|x\|^{-\varepsilon}$$

lie in finitely many proper linear subspaces of $\mathbb{Q}^n$.

This generalises the Thue-Siegel-Roth Theorem on the approximation of algebraic numbers [28] to higher dimensions.

Theorem 1.1 has been extended in various directions by many authors including Schmidt himself, Schlickewei, Evertse, Amoroso and Viada. Analogues have been proved using p-adic norms and over arbitrary number fields and bounds on the number of subspaces required have been found. These bounds depend on the degree of the number field and the dimension. For some of these results and more information see [20,21] and [2].

Now we give a p-adic version of the Subspace Theorem that we will use in the next section. Given a prime p, the *p-adic* absolute value is denoted $|x|_p$ and satisfies $|p|_p = 1/p$. $|x|_\infty$ denotes the usual absolute value so $|x|_\infty = |x|$. We may refer to ∞ as the *infinite prime*.

Theorem 1.2 (Subspace Theorem II). *Suppose* $S = \{\infty, p_1, \ldots, p_t\}$ *is a finite set of primes, including the infinite prime. For every* $p \in S$ *let* $L_{1,p}, \ldots, L_{n,p}$ *be linearly independent linear forms in n variables with algebraic coefficients. Then for any* $\varepsilon > 0$ *the solutions* $x \in \mathbb{Z}^n$ *of*

$$\prod_{p \in S} \prod_{i=1}^{n} |L_{i,p}(x)|_p \leq \|x\|^{-\varepsilon}$$

lie in finitely many proper linear subspaces of $\mathbb{Q}^n$.

In Theorem 1.2 we assume that a continuation of the p-adic norm to $\overline{\mathbb{Q}}$ has been chosen for each $p \in S$.

The power and utility of the Subspace Theorem is already evident in the above forms but there is a corollary which makes even more applications possible. This corollary was originally given by Evertse, Schlickewei and Schmidt [21]. We present the version with the best known bound due to Amoroso and Viada [2].

Theorem 1.3. *Let* K *be a field of characteristic* 0, Γ *a subgroup of* K^* *of rank* r, *and* $a_1, a_2, \ldots, a_n \in K^*$. *Then the number of solutions of the equation*

$$a_1 z_1 + a_2 z_2 + \cdots + a_n z_n = 1 \qquad (1.1)$$

with $z_i \in \Gamma$ *and no subsum on the left hand side vanishing is at most*

$$A(n, r) \leq (8n)^{4n^4(n+nr+1)}.$$

We will consider the following problems.

The Erdős unit distance problem is an important problem in combinatorial geometry. It asks for the maximum possible number of unit distances between n points in the plane. This problem is still open but recently Frank de Zeeuw and the authors have made progress towards this problem when the distances considered are roots of unity. Theorem 1.3 lets us consider distances from a group with rank not too large.

Given a finite set of real numbers we can define its sum set and product set. It is believed that for any such set either its sum set or product set is large. If the product set is "very" small then Theorem 1.3 gives that the sum set is "very" large.

There are many results in combinatorial geometry concerning the structure of lines. We highlight one such result about sets of lines with few intersection points.

The structure of this paper will be as follows. In the next section we give a number of well-known applications of Theorems 1.1-1.3. In Section 3 we give combinatorial applications and the special case of Theorem 1.3 via Mann's Theorem.

2 Number theoretic applications

In this section we give a few well-known applications of Theorems 1.1-1.3. These are not the best known results but they are given here with proofs as they may be of use to discrete mathematicians to illustrate how to use Theorems 1.1-1.3.

2.1 Transcendental numbers

Adamczewski and Bugeaud showed that all irrational automatic numbers are transcendental using the Subspace Theorem [1]. An *automatic number* is a number for which there exists a positive integer b such that when the number is written in b-ary form it is the output of a finite automaton with input the nonnegative integers written from right to left. For a detailed proof see the survey paper of Bilu [6].

Here we will use a method similar to the proof of Theorem 3.3 in [6] to show:

Theorem 2.1. *The number α given by the infinite sum*

$$\alpha = \sum_{n \geq 1} \frac{1}{2^{2^n}}$$

is transcendental.

Mahler showed in [26] that α is transcendental and Kempner showed that a large class of numbers defined similarly to α are transcendental [23]. The Subspace Theorem provides a tidy proof of this fact.

Proof of Theorem 2.1. Consider the binary expansion:

$$\alpha = \frac{1}{4} + \frac{1}{16} + \frac{1}{256} + \frac{1}{65536} + \cdots = 0.0101000100000001 \ldots {}_2 \, .$$

So the binary expansion of α consists of sections of zeros of increasing length separating solitary ones. Thus the expansion is not periodic and hence α is not rational. We let b_n be the string given by the first n digits of this expansion. One can check that each b_n has two disjoint substrings of zeros of length $n/8$.

Assume α is not transcendental. Then it is algebraic. Now each b_n starts with a string $AOBO$, where O is a string of zeroes, the length of O is at least $n/8$ and A and B might have length zero. We will use the rational number represented in base 2 by $0.AOBOBO \ldots$ to approximate α. Call this number π. Then

$$\pi = \frac{M}{2^a (2^b - 1)}$$

where $M \in \mathbb{Z}$ and a and b are the lengths of the strings A and OB respectively. Clearly $b \geq n/8$ and $a + b \leq n$ since AOB is a substring of b_n. Since α starts with b_n we have

$$|\alpha - \pi| \leq \frac{1}{2^{a+b+n/8}} \implies |2^{a+b}\alpha - 2^a\alpha - M| \leq \frac{1}{2^{n/8}}.$$

Now we apply the Subspace Theorem in the form given in Theorem 1.2. We let $S = \{2, \infty\}$ and

$$L_{1,\infty}(x) = x_1, \quad L_{2,\infty}(x) = x_2, \quad L_{3,\infty}(x) = \alpha x_1 - \alpha x_2 - x_3,$$
$$L_{1,2}(x) = x_1, \quad L_{2,2}(x) = x_2, \quad L_{3,2}(x) = x_3.$$

Note that by our assumption that α is not transcendental the linear form $L_{3,\infty}$ has algebraic coefficients. Let $x = (2^{a+b}, 2^a, M)$. Now $|M| \leq 2^{a+b}$ since $0 < \pi < 1$. So $\|x\| \leq 2^{a+b} \leq 2^n$. Multiplying the absolute values of the linear forms together we get

$$\prod_{p \in S} \prod_{i=1}^{3} |L_{i,p}(x)| = |2^a|_2 |2^a|_\infty |2^{a+b}|_2 |2^{a+b}|_\infty |M|_2 |2^{a+b}\alpha - 2^a\alpha - M|_\infty$$

$$\leq \frac{1}{2^{n/8}}$$

$$\leq \frac{1}{\|x\|^{1/8}}.$$

The first inequality holds because $|\alpha - \pi| \leq 2^{-a-b-n/8}$ and $|2|_2 |2|_\infty = 1$.

We can do this for each n and $b = b(n)$ increases as n increases since $b \geq n/8$. Thus infinitely many of the vectors $x = x(n)$ are distinct. By Theorem 1.2 these vectors are contained in finitely many subspaces of $\mathbb{Q}^3$. Thus one of these subspaces contains infinitely many of them. That is, there exist $c, d, e \in \mathbb{Q}$ such that

$$c2^{a(n)} + d2^{a(n)+b(n)} + eM(n) = 0$$

for infinitely many n. The coefficient e cannot be zero since $b(n) \to \infty$ as $n \to \infty$. Dividing by $2^{a(n)}(2^{b(n)} - 1)$ and taking limits we get $\alpha = -d/e$ so α is rational. This is a contradiction. Thus α must be transcendental. $\qquad \square$

2.2 Linear recurrence sequences

A linear recurrence sequence is a sequence of numbers where the first few terms are given and the higher order terms can be found by a recurrence relation. A famous example is the Fibonacci sequence $\{F_n\}$ where $F_0 = 0$, $F_1 = 1$ and $F_n = F_{n-1} + F_{n-2}$ for $n \geq 2$. More formally, a *linear recurrence sequence* consists of constants $a_1, \ldots, a_k$ in a field K for some $k > 0$ along with a sequence $\{R_m\}_{m=0}^{\infty}$ with $R_i \in K$ for $0 \leq i \leq k - 1$ and

$$R_n = a_1 R_{n-1} + a_2 R_{n-2} + \cdots + a_k R_{n-k}, \quad \text{for } n \geq k.$$

If $\{R_m\}$ is not expressible by any shorter recurrence relation then it is said to have order k. In this case $a_k \neq 0$.

We are interested in the structure of the zero set of a linear recurrence sequence. This is the set

$$S(\{R_m\}) = \{i \in \mathbb{N} : R_i = 0\}.$$

The Skolem-Mahler-Lech Theorem states that this set consists of the union of finitely many points and arithmetic progressions [25]. Schmidt has given a quantitative bound for this theorem using various tools including a variant of Theorem 1.3 [30].

We will show a special case of this theorem using Theorem 1.3. We will restrict our attention to simple nondegenerate linear recurrence sequences. To define such sequences we need to define the companion polynomial of the recurrence sequence. If $\{R_m\}$ is given as above then the *companion polynomial* of $\{R_m\}$ is $C(x) = x^k - a_1 x^{k-1} - \cdots - a_{k-1} x - a_k$. Suppose the roots of this polynomial are $\alpha_1, \ldots, \alpha_\ell$ with multiplicity $b_1, \ldots, b_\ell$ respectively. Each α_i is nonzero since $a_k \neq 0$. If the companion polynomial has k distinct roots it is called *simple*. If α_i / α_j is not a root of unity for any $i \neq j$ then the sequence is called *nondegenerate*. A version of this theorem was given in [20]. The improved bound given here is due to Amoroso and Viada [3].

Theorem 2.2. *Suppose $\{R_m\}$ is a simple nondegenerate linear recurrence sequence of order k with complex coefficients. Then*

$$|S(\{R_m\})| \leq (8k)^{8k^6}.$$

Proof. We can express the recurrence relation using a matrix equation:

$$\begin{bmatrix} a_1 & a_2 & \ldots & a_{k-1} & a_k \\ 1 & 0 & \ldots & 0 & 0 \\ 0 & 1 & \ldots & 0 & 0 \\ \vdots & \vdots & \ddots & \vdots & \vdots \\ 0 & 0 & \ldots & 1 & 0 \end{bmatrix}^n \begin{bmatrix} R_{k-1} \\ R_{k-2} \\ \vdots \\ R_0 \end{bmatrix} = \begin{bmatrix} R_{k-1+n} \\ R_{k-2+n} \\ \vdots \\ R_n \end{bmatrix}. \tag{2.1}$$

We call the $k \times k$ matrix above A. The characteristic polynomial of A is given by

$$\chi(\lambda) = (-1)^k (\lambda^k - a_1 \lambda^{k-1} - \cdots - a_k).$$

This is the same, up to sign, as the companion polynomial of $\{R_m\}$. Thus A has distinct nonzero eigenvalues and so can be diagonalized. So $A = PDP^{-1}$ for some invertible $k \times k$ matrix P and

$$D = \begin{bmatrix} \alpha_1 & 0 & \cdots & 0 \\ 0 & \alpha_2 & \cdots & 0 \\ \vdots & \vdots & \ddots & 0 \\ 0 & 0 & \cdots & \alpha_k \end{bmatrix}.$$

So D^n is a diagonal matrix with α_i^n in the i-th row and column. Thus, multiplying out $PD^n P^{-1}$ we see that every element of A^n is a linear combination of the n-th powers of the α_i's. Now, by the matrix equation in (2.1) we see that R_n is given by the k-th row of A^n multiplied by $[R_{k-1}, R_{k-2}, \ldots, R_0]^T$ and so

$$R_n = c_1 \alpha_1^n + c_2 \alpha_2^n + \cdots + c_k \alpha_k^n \quad \text{for every } n \geq k.$$

Then applying Theorem 1.3 to the equation $c_1 x_1 + c_2 x_2 + \cdots + c_k x_k = 0$ with solutions from the group of rank at most k generated by $\{\alpha_1, \ldots, \alpha_k\}$ we get that the number of solutions is at most

$$A(k, k) \leq (8k)^{4k^4(k+k^2+1)} \leq (8k)^{8k^6}.$$

Since the sequence is nondegenerate we cannot have two values n, n' giving the same value for α_i^n and $\alpha_i^{n'}$ for each i, hence each solution corresponds to a unique value from $S(\{R_m\})$.   □

3 Combinatorial applications

3.1 A proof of a very special case of Theorem 1.3

Theorem 1.3 gives a bound on the number of nondegenerate solutions of a linear equation from a multiplicative group with rank not too large. What happens if the group in question has rank zero? This corresponds to solutions that are roots of unity. Theorem 1.3 can then be seen as a generalisation of the following result which follows from a theorem of H.B. Mann from 1965.

Theorem 3.1. *Given* $(a_1, \ldots, a_k) \in \mathbb{Q}^k$, $a \in \mathbb{C}^*$, *consider the equation*

$$a_1 x_1 + a_2 x_2 + \cdots + a_k x_k = a.$$

The number of solutions $(\omega_1, \ldots, \omega_k)$ of this equation with the ω_i's roots of unity and no vanishing subsum is at most $(k \cdot \Theta(2k))^k$ where

$$\Theta(k) = \prod_{\substack{p \leq k \\ p \text{ prime}}} p.$$

Note that the logarithm of the function Θ above is an important function in number theory called the first Chebyshev function.

Theorem 3.1 along with Lemma 3.2 below were proved by Frank de Zeeuw and the authors in [32]. The roots of unity give a relatively simple example of an infinite multiplicative group. We will give the proof of Theorem 3.1 below. First we prove Lemma 3.2 which was Mann's result mentioned above [27].

Lemma 3.2 (Mann). *Suppose we have*

$$a_1\omega_1 + a_2\omega_2 + \cdots + a_k\omega_k = 0,$$

with $a_i \in \mathbb{Q}$, the ω_i's roots of unity, and no proper nontrivial subsum vanishing. Then for every $1 \leq i, j \leq k$, $(\omega_i/\omega_j)^{\Theta(k)} = 1$.

A result of Conway and Jones gives an improved bound in Lemma 3.2 and hence in Theorem 3.1 [10]. Evertse has also given a bound on sums of the form given in Theorem 3.1 but with $a_1, \ldots, a_k \in \mathbb{C}^*$ [19].

Proof. Dividing by an appropriate factor we may assume that our equation is of the form $1 + a_2\omega_2 + \cdots + a_k\omega_k = 0$. Then we just need to show that each $\omega_i^{\Theta(k)} = 1$. We let s be the smallest value such that $\omega_i^s = 1$ for $1 \leq i \leq k$. The proof proceeds by showing that s is squarefree and any prime that divides s cannot be larger than k. This means $s \leq \Theta(k)$.

Suppose p^j divides s exactly. Then each $\omega_i = \rho^{\sigma_i} \cdot \omega_i^*$, with ρ a primitive p^j-th root of unity, $\sigma_i < p$ and ω_i^* an (s/p)-th root of unity. So rewriting the sum grouping powers of ρ we get

$$0 = 1 + (a_2\omega_2 + \cdots + a_k\omega_k) = 1 + (\alpha_0 + \alpha_1\rho + \cdots + \alpha_{p-1}\rho^{p-1}),$$

where, for each i, $\alpha_i \in K := \mathbb{Q}(\omega_2^*, \ldots, \omega_k^*)$ satisfies

$$\alpha_\ell = \sum_{i \in I_\ell} a_i\omega_i^*, \text{ with } I_\ell = \{i : \sigma_i = \ell\}.$$

Let $f(x) = \alpha_{p-1}x^{p-1} + \cdots + \alpha_1 x + (1 + \alpha_0)$. Then f is a polynomial of degree at most $p - 1$ over the field K and $f(\rho) = 0$. If f were identically zero then, by the minimality of s, we would have a vanishing subsum.

The degree of ρ over K gives us that p divides s only once. Specifically since $[K(\rho) : \mathbb{Q}] = [K(\rho) : K][K : \mathbb{Q}]$ we have

$$\deg_K(\rho) = [K(\rho) : K] = \frac{[K(\rho) : \mathbb{Q}]}{[K : \mathbb{Q}]} = \frac{\phi(s)}{\phi(s/p)}.$$

This is p if $j > 1$ and $p - 1$ if $j = 1$. But the degree of f is at most $p - 1$ so we must have $j = 1$ since ρ is a root of f.

Now f must be a multiple of the irreducible polynomial m of ρ over K. But $m(x) = x^{p-1} + x^{p-2} + \cdots + 1$ so $f(x) = cm(x)$ where c is a nonzero constant. Thus f has p nonzero coefficients and thus so does the original sum giving $p \leq k$.   □

Proof of Theorem 3.1. We first show that if we are given $a \in \mathbb{C}^*$ and two sums $a_1\omega_1 + \cdots + a_k\omega_k = a$ and $a_1'\omega_1' + \cdots + a_k'\omega_k' = a$ with rational coefficients and no vanishing subsums then for any ω_j', there is an ω_i such that $(\omega_j'/\omega_i)^{\Theta(2k)} = 1$.

Since $a_1\omega_1 + \cdots + a_k\omega_k = a = a_1'\omega_1' + \cdots + a_k'\omega_k'$, we get

$$a_1\omega_1 + \cdots + a_k\omega_k - a_1'\omega_1' - \cdots - a_k'\omega_k' = 0. \qquad (3.1)$$

This sum may have vanishing subsums so we consider minimal vanishing subsums of the form

$$\sum_{i \in I_\ell} a_i\omega_i - \sum_{j \in I_\ell'} a_j'\omega_j' = 0.$$

Each ω_j' is contained in such a minimal subsum of length at most $2k$. This subsum also contains some ω_i otherwise the original sum would have a vanishing subsum. Now the previous lemma gives that $(\omega_j'/\omega_i)^{\Theta(2k)} = 1$.

Note that above we require $a \in \mathbb{C}^*$. If $a = 0$ then the two original sums will count as vanishing subsums when we consider the combined equation (3.1) so Lemma 3.2 does not apply.

Now we can prove the theorem. For $a \in \mathbb{C}^*$ and k a positive integer define $S(a, k)$ as the set of k-tuples $(\omega_1, \ldots, \omega_k)$, where each ω_i is a root of unity, such that there are $a_i \in \mathbb{Q}$ satisfying $a_1\omega_1 + \cdots + a_k\omega_k = a$ with no vanishing subsums.

We fix a k-tuple $(\omega_1, \ldots, \omega_k) \in S(a, k)$. Given an element of $S(a, k)$, for each ω_j' (the j-th coordinate of that element) there is an i such that $\omega_i^{-\Theta(2k)}(\omega_j')^{\Theta(2k)} = 1$. So ω_j' is a root of the polynomial $\omega_i^{-\Theta(2k)}x^{\Theta(2k)} = 1$. This polynomial has $\Theta(2k)$ roots. We have k choices for j so at most $k\Theta(2k)$ choices for each ω_j'. This gives the required bound.   □

This theorem can be used to prove Theorem 3.3 from the next section. We will show how using Theorem 1.3 instead allows the proof of the stronger Theorem 3.4.

3.2   Unit distances

The unit distance problem was first posed by Erdős in 1946 [17]. It asks for the maximum number, $u(n)$, of pairs of points with the same distance in a collection of n points in the plane. By scaling the point set one may assume that the most popular distance is one, hence the name of the problem. The problem seeks asymptotic bounds. Erdős gave a construction using a $\sqrt{n} \times \sqrt{n}$ portion of a square lattice giving

$$u(n) \geq n^{1+c/\log\log n}.$$

Number theoretic bounds for the number of integer solutions of the equation $x^2 + y^2 = a$ give the above inequality. Erdős conjectured that the magnitude of $u(n)$ is close to this lower bound. The best known upper bound is $u(n) \leq cn^{4/3}$. A number of proofs have been given showing $u(n) \leq cn^{4/3}$ using tools such as cuttings, edge crossings in graphs and the Szemerédi-Trotter Theorem. The first proof was due to Spencer, Szemerédi and Trotter [35]. For more details of the problem see [7]. We will look at a special case of this problem when the distances considered come from a multiplicative group with rank not too large. This does not seem to be a huge limitation as the unit distances from the lower bound construction above come from such a group as will be explained below.

Using Theorem 3.1 Frank de Zeeuw and the authors were able to show the following theorem [32]. Two points in the plane are said to have *rational angle* if the angle that the line between these two points makes with the x-axis is a rational multiple of π.

Theorem 3.3. *Let $\varepsilon > 0$. Given n points in the plane, the number of unit distances with rational angle between pairs of points is less than $n^{1+\varepsilon}$.*

These unit distances correspond to roots of unity. The proof proceeds by counting certain paths in the unit distance graph and using Mann's Theorem to bound the number of edges.

Using Theorem 1.3 in place of Mann's Theorem one can instead consider unit distances from a multiplicative group with rank not too large with respect to the number of points [31]. Note that a unit distance in the plane can (and will) be considered as a complex number of unit length. So all unit distances can be considered as coming from a subgroup of $\mathbb{C}^*$.

Theorem 3.4. *Let $\varepsilon > 0$. There exist a positive integer n_0 and a constant $c > 0$ such that given $n > n_0$ points in the plane, the number of unit distances coming from a subgroup $\Gamma \subset \mathbb{C}^*$ with rank $r < c\log n$ is less than $n^{1+\varepsilon}$.*

This is our first combinatorial application of Theorem 1.3. The proof is given below.

Suppose $G = G(V, E)$ is a graph on $v(G) = n$ vertices and $e(G) = m$ edges. We denote the minimum degree in G by $\delta(G)$.

Note that by removing vertices with degree less than $m/(2n)$ we have a subgraph H with at least $e(H) \geq m/2$ edges and $\delta(H) \geq m/(2n)$. The number of vertices in H is at least $v(H) \geq \sqrt{m}$. We will consider such a well behaved subgraph instead of the original graph.

Proof of Theorem 3.4. Let G be the unit distance graph on n points with unit distances coming from Γ as edges. We show that there are less than $n^{1+\varepsilon}$ such edges, i.e. distances, for any $\varepsilon > 0$. We can assume that $e(G) \geq (1/2)n^{1+\varepsilon}$, $v(G) \geq n^{1/2+\varepsilon/2}$ and $\delta(G) \geq (1/2)n^{\varepsilon}$.

Consider a path in G on k edges $P_k = p_0 p_1 \ldots p_k$. We denote by $u_i(P_k)$ the unit vector between p_i and p_{i+1}. The path is *nondegenerate* if $\sum_{i \in I} u_i(P_k) = 0$ has no solutions where I is a nonempty subset of $\{0, 1, \ldots, k-1\}$. Note that such a sum is a sum of elements of Γ with no vanishing subsums. We will denote by $\mathcal{P}_k(v, w)$ the set of nondegenerate paths of length k between vertices v and w.

The number of nondegenerate paths of length k from any vertex is at least

$$\prod_{\ell=0}^{k-1} (\delta(G) - 2^{\ell} + 1) \geq \frac{n^{k\varepsilon}}{2^{2k}}.$$

The first expression is true since if we consider a path P_ℓ on $\ell < k$ edges then all but $2^\ell - 1$ possible continuations give a path $P_{\ell+1}$ with no vanishing subsums. The inequality is true if we assume $2^k \leq (1/2)n^\varepsilon$, which is true if $k < (\varepsilon \log n)/\log 2 - 1$, a fact we will confirm at the end of the proof. From this we get that the number of nondegenerate paths P_k in the graph is at least $n^{1/2+(k+1/2)\varepsilon}/2^{2k+1}$. So there exist vertices v, w in G with

$$|\mathcal{P}_k(v, w)| \geq \frac{n^{(k+1/2)\varepsilon - 3/2}}{4^k}.$$

Consider a path $P_k \in \mathcal{P}_k(v, w)$, $P_k = p_0 p_1 \ldots p_k$. Let a be the complex number giving the vector between p_0 and p_k. Since P_k is nondegenerate we get a solution of $(1/a)x_1 + (1/a)x_2 + \cdots + (1/a)x_k = 1$ with no vanishing subsums. Thus Theorem 1.3 gives

$$|\mathcal{P}_k(v, w)| \leq (8k)^{4k^4(k+kr+1)}.$$

This with the lower bound give

$$((k + 1/2)\varepsilon - 3/2) \log n \le k \log 4 + 4k^4(k + kr + 1) \log(8k)$$
$$\le c'rk^5 \log k,$$
$$\implies \varepsilon \le \frac{c'rk^4 \log k}{\log n} + \frac{c''}{k}. \tag{3.2}$$

Since $r + 1 \le c \log n$ we can choose k an integer satisfying

$$C'((\log n)/r)^{1/5} \le k \le C''((\log n)/r)^{1/5}.$$

Then, with this k, the right hand side of (3.2) goes to zero as n increases. Earlier we assumed that $k \le (\varepsilon \log n)/\log 2 - 1$. This holds for the value of k given above for n large enough. So the number of unit distances from Γ is less than $cn^{1+\varepsilon}$ for each $\varepsilon > 0$. $\qquad\square$

Performing a careful analysis of Erdős' lower bound construction one can show that all unit distances come from a group with rank at most $c \log n / \log \log n$ for some $c > 0$. This group is generated by considering solutions of the equation $x^2 + y^2 = p$ where p is a prime of the form $4m + 1$. Using the prime number theorem for arithmetic progressions we get the bound on such solutions and thus on the rank. For all the details see [31]. So Erdős' construction satisfies the conditions of Theorem 3.4. A similar approach could be used for other types of lattices. So all the best known lower bounds for the unit distance problem have unit distances coming from a well structured group. It would be interesting to see if any configuration of points with maximum unit distances has such a structure.

3.3 Sum-product estimates

The theory of sum sets and product sets plays an important part in combinatorics and additive number theory. The goal of the field is to show that for any finite subset of a field either the sum set or the product set is large. We will focus on the complex numbers.

Formally, given a set $A \subset \mathbb{C}$, the sum set, denoted by $A + A$, and product set, denoted by AA, are

$$A + A := \{a + b : a, b \in A\}, \qquad AA := \{ab : a, b \in A\}.$$

Note that

$$|A| \le |A + A|, |AA| \le \binom{|A| + 1}{2} = \frac{|A|^2}{2} + \frac{|A|}{2}.$$

The following long standing conjecture of Erdős and Szemerédi [18] has led to much work in the field.

Conjecture 3.5. Let $\varepsilon > 0$ and $A \subset \mathbb{Z}$ with $|A| = n$. Then

$$|A + A| + |AA| \geq Cn^{2-\varepsilon}.$$

This conjecture is still out of reach. The best known bound, which holds for real numbers and not just integers, is $Cn^{4/3-o(1)}$ due to Solymosi [34]. A similar bound was proved recently by Konyagin and Rudnev in [24].

Chang showed that when the product set is small Theorem 1.3 can be used to show that the sum set is large [8]. The following reformulation of Chang's observation is due to Andrew Granville.

Theorem 3.6. *Let $A \subset \mathbb{C}$ with $|A| = n$. Suppose $|AA| \leq Cn$. Then there is a constant C' depending only on C such that*

$$|A + A| \geq \frac{n^2}{2} + C'n.$$

We will present a proof of Theorem 3.6 below. To use Theorem 1.3 we need a multiplicative subgroup with finite rank to work with. The following lemma of Freiman, which appears as Lemma 1.14 in [22], provides this.

Lemma 3.7 (Freiman). *Let $A \subset \mathbb{C}$. If $|AA| \leq C|A|$ then A is a subset of a multiplicative subgroup of $\mathbb{C}^*$ of rank at most r, where r is a constant depending on C.*

Proof of Theorem 3.6. We consider solutions of $x_1 + x_2 = x_3 + x_4$ with $x_i \in A$. A solution of this equation corresponds to two pairs of elements from A that give the same element in $A+A$. Let us suppose that $x_1+x_2 \neq 0$ (there are at most $|A| = n$ solutions of the equation $x_1 + x_2 = 0$ with $x_1, x_2 \in A$.)

First we consider the solutions with $x_4 = 0$. Then by rearranging we get

$$\frac{x_1}{x_3} + \frac{x_2}{x_3} = 1. \tag{3.3}$$

By Lemma 3.7 and Theorem 1.3 there are at most $s_1(C)$ solutions of $y_1 + y_2 = 1$ with no subsum vanishing. Each of these gives at most n solutions of (3.3) since there are n choices for x_3. There are only two solutions of $y_1 + y_2 = 1$ with a vanishing subsum, namely $y_1 = 0$ or $y_2 = 0$, and each of these gives n solutions of (3.3). So we have a total of $(s_1(C) + 2)n$ solutions of (3.3).

For $x_4 \neq 0$ we get

$$\frac{x_1}{x_4} + \frac{x_2}{x_4} - \frac{x_3}{x_4} = 1. \tag{3.4}$$

Again by Freiman's Lemma and Theorem 1.3, the number of solutions of this with no vanishing subsum is at most $s_2(C)n$. If we have a vanishing subsum then $x_1 = -x_2$ which is a case we excluded earlier or $x_1 = x_3$ and then $x_2 = x_4$, or $x_2 = x_3$ and then $x_1 = x_4$. So we get at most $2n^2$ solutions of (3.4) with a vanishing subsum (these are the $x_1 + x_2 = x_2 + x_1$ identities.)

So, in total, we have at most $2n^2 + s(C)n$ solutions of $x_1 + x_2 = x_3 + x_4$ with $x_i \in A$. Suppose $|A + A| = k$ and $A + A = \{\alpha_1, \ldots, \alpha_k\}$. We may assume that $\alpha_1 = 0$. Recall that we ignore sums $a_i + a_j = 0$. Let

$$P_i = \{(a, b) \in A \times A : a + b = \alpha_i\}, \quad 2 \le i \le k.$$

Then

$$\sum_{i=2}^{k} |P_i| \ge n^2 - n = n(n - 1).$$

Also, a solution of $x_1 + x_2 = x_3 + x_4$ corresponds to picking two values from P_i where $x_1 + x_2 = \alpha_i$. Thus

$$2n^2 + s(C)n \ge \sum_{i=2}^{k} |P_i|^2 \ge \frac{1}{k-1} \left(\sum_{i=2}^{k} |P_i| \right)^2 \ge \frac{n^2(n-1)^2}{k-1}$$

by the Cauchy-Schwarz inequality. The bound for k follows. $\square$

Note that in this paper we use a simple bound, Freiman's lemma, on the rank of the multiplicative group. Results in the direction of the so called polynomial Freiman conjecture give better bounds on the rank of a large subset of the set with small product set.

3.4   Line configurations with few intersections

A number of other combinatorial results follows from Theorem 1.3. We give one more of these, from combinatorial geometry. This is similar to a result due to Chang and Solymosi [9]. A *complex line* is a line in the complex plane. Specifically, a complex line is given by an equation $ax + by = c$ where $a, b, c \in \mathbb{C}$ and x and y are the (complex) coordinates in $\mathbb{C}^2$. Given two lines L and M we denote their intersection point by $L \cap M$.

Theorem 3.8. *Let $C > 0$. Then there exists $c > 0$ such that for any $n + 3$ lines $L_1, L_2, L_3, M_1, \ldots, M_n$ in $\mathbb{C}^2$, with the L_i not all parallel and $L_1 \cap L_2, L_1 \cap L_3$ and $L_2 \cap L_3$ distinct the following holds. If the number of distinct intersection points $L_i \cap M_j, 1 \le i \le 3, 1 \le j \le n$, is at most $C\sqrt{n}$ then any line $L \notin \{L_1, L_2, L_3\}$ has at least cn distinct intersection points $L \cap M_j, 1 \le j \le n$.*

There are many structure results similar to Theorem 3.8 in discrete geometry. These include Beck's Theorem [5], a structure theorem for lines containing many points of a cartesian product by Elekes [14] and generalisations of this line theorem to surfaces by Elekes and Rónyai [15], Elekes and Szabó [16] and Frank de Zeeuw and the authors [33]. The proofs of these results used the Szemerédi-Trotter Theorem and techniques from commutative algebra and algebraic geometry. These theorems have been used to prove various results including a conjecture of Purdy about the number of distinct distances between two sets of collinear points in the plane. For more details see [12, 13] and [33].

We do not prove Theorem 3.8 completely but only give a sketch of how it follows from Theorem 1.3. We don't try to find an efficient quantitative version here and we don't explain the methods used in detail. The techniques applied are standard methods in additive combinatorics. All the details can be found in [36]. The proof requires a refinement of the Balog-Szemerédi Theorem [4]. We are going to use the notation of *sums along a graph*. For two subsets of a group A and B and a bipartite graph $G = G(A, B)$ with vertex classes A and B the sums (or products) along G is the set

$$A +_G B = \{a + b | a \in A, b \in B, (a, b) \in G\}.$$

The cardinality of G, which is denoted by $|G|$, is the number of edges in G.

Theorem 3.9 (Balog-Szemerédi). *Let us suppose that A and B are two finite subsets of an abelian group such that $|A| = |B| = m$ and $|A +_G B| \leq Cm$ where $|G(A, B)| \geq cm^2$. Then there are sets $A' \in A, B' \in B$ such that $|A' + B'| \leq Dm$ and $|A' \times B' \cap G(A, B)| \geq \delta m^2$ where D and $\delta > 0$ depends on c and C only.*

Apply a projective transformation which moves L_1 to the x-axis, L_2 to the y-axis, and L_3 to the horizontal line $y = 1$. The three lines have distinct intersection points thus such a transformation exists. Let us denote the x-coordinates of $L_1 \cap M_i$ and $L_3 \cap M_j$ by x_i and y_j respectively. The two sets of x-coordinates are denoted by X and Y. Define a bipartite graph with vertices given by the x-coordinates of the intersection points of the lines M_i with L_1 and L_3 (with vertex sets X and Y without multiplicity.) Two points are connected by an edge in the graph if they are connected by a line M_j. This is a bipartite graph on at most $C\sqrt{n}$ vertices with n edges. If $M_i \cap L_2$ is the point $(0, \alpha)$ then $x_i/y_i = \alpha/(\alpha - 1)$, or equivalently $x_i = \alpha y_i/(\alpha - 1)$. The Balog-Szemerédi Theorem and Freiman's Lemma imply that there are large subsets $X' \subset V_1$ and $Y' \subset V_2$ so that X' and Y'

are subsets of a multiplicative subgroup of $\mathbb{C}^*$ of rank at most $r(C)$ and the subgraph spanned by X', Y' still has at least some δn edges. We show that the lines represented by these δn edges cannot have high multiplicity intersections outside of L_1, L_2, L_3. If (a, b) is a point of M_i connecting two points $x_i \in X'$ and $y_i \in Y'$ then $(a - x_i)/(a - y_i) = b/(b-1)$, which gives a solution (x_i, y_i) to the equation $cx + dy = 1$ if $a \neq 0, b \neq 0, 1$. Here c, d depend on a and b only. As x_i and y_i are from a multiplicative group of bounded rank, we have a uniform bound, B, on the number of lines between X' and Y' which are incident to (a, b). There are δn lines connecting at most $C\sqrt{n}$ points. No more than $C\sqrt{n}/2$ of them might be parallel to any given line. Any line intersects at least $\delta n - C\sqrt{n}$ of them. Any intersection point outside of the lines L_1, L_2, and L_3 is incident to at most B lines, so there are at least cn distinct intersection points $L \cap M_j, 1 \leq j \leq n$, with any other line.

We are unaware of any proof of this fact without Theorem 1.3.

ACKNOWLEDGEMENTS. The work in Section 3.1 is joint work with Frank de Zeeuw. The authors are thankful to Jarik Nešetřil for the encouragement to write this survey. We are also thankful to the organizers of the workshop in Pisa, "Geometry, Structure and Randomness in Combinatorics", where the parts of this paper were presented. The authors would also like to thank the anonymous referee for their helpful comments and corrections.

References

[1] B. ADAMCZEWSKI and Y. BUGEAUD, *On the complexity of algebraic numbers I: Expansions in integer bases*, Annals of Mathematics **165** (2007), 547–565.

[2] F. AMOROSO and E. VIADA, *Small points on subvarieties of a torus*, Duke Mathematical Journal, **150**(3) (2009), 407–442.

[3] F. AMOROSO and E. VIADA, *On the zeros of linear recurrence sequences*, Acta Arithmetica **147** (2011), 387–396.

[4] A. BALOG and E. SZEMERÉDI, *A statistical theorem of set addition*, Combinatorica **14** (1994), 263–268.

[5] J. BECK, *On the lattice property of the plane and some problems of Dirac, Motzkin, and Erdős in combinatorial geometry*, Combinatorica **3**(3) (1983), 281–297.

[6] Y. BILU, *The Many Faces of the Subspace Theorem (after Adamczewski, Bugeaud, Corvaja, Zannier...)*, Séminaire Bourbaki, Ex-

posé 967, 59ème année (2006-2007); Astérisque **317** (2008), 1–38., May 2007.

[7] P. BRASS, W. MOSER and J. PACH, "Research Problems in Discrete Geometry", chapter 5: Distance Problems, Springer, 2006, 183–257.

[8] M.-C. CHANG, *Sum and product of different sets*, Contributions to Discrete Mathematics **1**(1), 2006.

[9] M.-C. CHANG, *Sum-product theorems and incidence geometry*, Journal of the European Mathematical Society **9**(3) (2007), 545–560.

[10] J. H. CONWAY and A. J. JONES, *Trigonometric diophantine equations (on vanishing sums of roots of unity)*, Acta Arithmetica **30** (1976), 229–240.

[11] P. CORVAJA and U. ZANNIER, *Applications of the subspace theorem to certain diophantine problems*, In: H. P. Schlickewei, Schmidt K., and R. F. Tichy (eds.), "Diophantine Approximation", Developments in Mathematics, Vol. 16, Springer Vienna, 2008, 161–174.

[12] G. ELEKES, *A note on the number of distinct distances*, Periodica Mathematica Hungarica **38**(3) (1999), 173–177.

[13] G. ELEKES *Sums versus products in number theory, algebra and Erdős geometry*, In: "Paul Erdős and his Mathematics II", *Bolyai Society Mathematical Studies*, Vol. 11, 2002, 241–290.

[14] GY. ELEKES, *On linear combinatorics, I*, Combinatorica **17**(4) (1997), 447–458.

[15] GY. ELEKES and L. RÓNYAI, *A combinatorial problem on polynomials and rational functions*, Journal of Combinatorial Theory, Series A **89** (2000), 1–20.

[16] GY. ELEKES and E. SZABÓ, *How to find groups? (and how to use them in Erdős geometry?)*, Combinatorica (2012), 1–35, 10.1007/s00493-012-2505-6.

[17] P. ERDŐS, *On sets of distances of n points*, American Mathematical Monthly **53**(5), 248–250, May 1946.

[18] P. ERDŐS and E. SZEMERÉDI, *On sums and products of integers*, In: "Studies in Pure Mathematics", Birkhäuser, 1983, 213–218.

[19] J.-H. EVERTSE, *The number of solutions of linear equations in roots of unity*, Acta Arithmetica **89**(1), 1999.

[20] J.-H. EVERTSE and H. P. SCHLICKEWEI, *The absolute subspace theorem and linear equations with unknowns from a multiplicative group*, In: K. Györy, H. Iwaniec, and J. Urbanowicz (eds.), "Number theory in progress: Proceedings of the international conference of number theory in honour of the 60th birthday of Andrzej Schinzel", 1999.

[21] J.-H. EVERTSE, H. P. SCHLICKEWEI and W. M. SCHMIDT, *Linear equations in variables which lie in a multiplicative group*, Annals of Mathematics **155**(3) (2002), 807–836.

[22] G. A. FREIMAN, "Foundations of a Structural Theory of Set Addition" Translations of Mathematical Monographs. American Mathematical Society, 1973.

[23] A. J. KEMPNER, *On transcendental numbers*, Transactions of the American Mathematical Society **17**(4) (1916), 476–482.

[24] S. V. KONYAGIN and M. RUDNEV, *On new sum-product type estimates*, Preprint, arXiv:1207.6785, 2013.

[25] C. LECH, *A note on recurring series*, Arkiv der Mathematik **2** (1953), 417–421.

[26] K. MAHLER, Über einen Satz von Mellin. *Mathematische Annalen*, 101:342–366, 1929.

[27] H. B. MANN, *On linear relations between roots of unity*, Mathematika **12** (1965), 107–117.

[28] K. F. ROTH, *Rational approximations to algebraic numbers*, Mathematika **2**(1) (1955), 1–20.

[29] W. M. SCHMIDT, *Norm form equations*, Annals of Mathematics **96**(3) (1972), 526–551.

[30] W. M. SCHMIDT, *The zero multiplicity of linear recurrence sequences*, Acta Mathematica **182**(2) (1999), 243–282.

[31] R. SCHWARTZ, *Using the subspace theorem to bound unit distances*, Moscow Journal of Combinatorics and Number Theory **3**(1) (2013), 108–117.

[32] R. SCHWARTZ, J. SOLYMOSI and F. DE ZEEUW, *Rational distances with rational angles*, Mathematika **58**(2) (2012), 409–418.

[33] R. SCHWARTZ, J. SOLYMOSI and F. DE ZEEUW, *Extensions of a result of Elekes and Rónyai*, Journal of Combinatorial Theory, Series A, 120(7) (2013), 1695–1713.

[34] J. SOLYMOSI, *Bounding multiplicative energy by the sumset*, Advances in Mathematics **222**(2) (2009), 402–408.

[35] J. SPENCER, E. SZEMERÉDI and W. TROTTER, *Unit distances in the Euclidean plane*, In: B. Bollobas (ed.) "Graph Theory and Combinatorics: Proceedings of the Cambridge Combinatorial Conference", in Honour of Paul Erdős, Academic Press, 1984, 293–303.

[36] T. TAO and V. H. VU, "Additive Combinatorics" Cambridge Studies in Advanced Mathematics 105. Cambridge University Press, 2006.

Can connected commuting graphs of finite groups have arbitrarily large diameter?

Peter Hegarty and Dmitry Zhelezov

Abstract. We present a two-parameter family, of finite, non-abelian random groups and propose that, for each fixed k, as $m \to \infty$ the commuting graph of $G_{m,k}$ is almost surely connected and of diameter k. As well as being of independent interest, our groups would, if our conjecture is true, provide a large family of counterexamples to the conjecture of Iranmanesh and Jafarzadeh that the commuting graph of a finite group, if connected, must have a bounded diameter.

We present a way to construct a family of random groups related to the conjecture of Iranmanesh and Jafarzadeh about commuting graphs of finite groups. Let G be a non-abelian group. We define the *commuting graph* of G, denoted by $\Gamma(G)$, as the graph whose vertices are the non-central elements of G, and such that $\{x, y\}$ is an edge if and only if $xy = yx$. One can just as well define the graph to have as its vertices the non-identity cosets of $Z(G)$, with $\{Zx, Zy\}$ adjacent if and only if $xy = yx$ and we stick to this definition henceforth. The conjecture of Iranmanesh and Jafarzadeh is as follows.

Conjecture 1. (Iranmanesh and Jafarzadeh, [5]) There is a natural number b such that if G is a finite, non-abelian group with $\Gamma(G)$ connected, then $\mathrm{diam}(\Gamma(G)) \leq b$.

The initial motivation was to show that Conjecture 1 is false by providing a counterexample using probabilistic methods. Some partial results in favor of Conjecture 1 (see details in the full length version of the present note, [4]) were already known at the moment the work on this project was initiated. It might seem natural to guess that for the commuting graph to be of large diameter, the group itself should be far from being abelian. However, it turns out in many cases the opposite holds and the commuting graph is connected and is of small diameter. It is thus reasonable to look at "more abelian" groups. Guidici and Pope [3] were first to consider the case of p-groups and provided a few notable results in support of Conjecture 1.

Let us recall some basic definitions first. If x, y are two elements of a group G, then their *commutator* $[x, y]$ is defined to be the group element

$x^{-1}y^{-1}xy$. The commutator subgroup of G is the subgroup generated by all the commutators and is denoted G'. If $G' \subseteq Z(G)$ one says that G is of *nilpotence class* 2. Quite surprisingly, one of the results of Guidici and Pope was that in this case the centre of the group should be of considerable size, otherwise the conjecture holds.

Theorem 2. *If G is of nilpotence class* 2 *and* $|Z(G)|^3 < |G|$, *then* $diam(\Gamma(G)) = 2$.

The general idea behind our construction is that if Conjecture 1 is false, then it should already fail among groups of nilpotence class two. Even more, one can take G such that both $Z(G)$ and $G/Z(G)$ are both elementary abelian 2-groups, that is, additive groups of some vector spaces over $\mathbb{F}_2$. However, instead of trying to construct an explicit counterexample we are going to introduce randomness in defining commutator relations in order to study how the commuting graph of a typical group of that kind looks like. As illustrated by many applications of the probabilistic method pioneered by Erdős (see [1] for the full treatment), the behaviour of a random object is often easer to analyse, so by adjusting parameters it is sometimes possible to provide an example with desired properties. Unfortunately, we were unable to disprove the conjecture in full in this way, but were able to produce a group whose commuting graph is of diameter 10, which became the largest value achieved by that time[1].

Before we proceed with the model of random groups, let us describe the significant success which took place since our work was undertaken. In [2], Giudici and Parker provide explicit examples of connected commuting graphs of unbounded diameter, thus disproving Conjecture 1. Their construction is based on and inspired by the random groups presented here, though they were able to devise an explicit construction. They have checked by computer that their model produces examples of commuting graphs of every diameter between 3 and 15, though it appears to remain open whether every positive integer diameter is achievable. As a remarkable counterpoint to their result, Morgan and Parker [6] have proved that if G has trivial centre then every connected component of $\Gamma(G)$ has diameter at most 10. Note that this condition specifically excludes nilpotent groups. In contrast to these purely group-theoretical advances, we are not aware of any further progress having been made on the analysis of the random groups described below.

Returning to our random construction, the group is defined as follows. Let m, r be positive integers and $V = V_m$ and $H = H_r$ be vector spaces

---

[1] September 2012.

over $\mathbb{F}_2$ of dimensions m and r respectively. Let $\phi : V \times V \to H$ be a bilinear map. Set $G := V \times H$ and define a multiplication on G by

$$(v_1, h_1) \cdot (v_2, h_2) := (v_1 + v_2, h_1 + h_2 + \phi(v_1, v_2)). \tag{1}$$

Then it is easy to check that

(i) $(G, \cdot)$ is a group of order 2^{m+r}, with identity element $(0, 0)$.
(ii) Let $\mathcal{H} := \{(0, h) : h \in H\}$. Then $\mathcal{H}$ is a subgroup of G and $G/\mathcal{H} \cong V$, as an abelian group.
(iii) $G' \subseteq \mathcal{H} \subseteq Z(G)$.
(iv) G is abelian if and only if ϕ is symmetric.
(v) The commutator of two elements is given by

$$[(v_1, h_1), (v_2, h_2)] = (0, \phi(v_1, v_2) - \phi(v_2, v_1)) \tag{2}$$

The map $\phi(\cdot, \cdot)$ is taken uniformly at random among all possible bilinear maps. It is then clear, due to (2), that, for two fixed distinct elements of G, their commutator becomes uniformly distributed on $\mathcal{H}$. Moreover, if we fix a basis $(v_1, ..., v_m)$ of V then all the commutator relations are determined by the skew-symmetric matrix A with $A_{i,j} = \phi(v_i, v_j) - \phi(v_j, v_i)$. Now we are going to define the parameters m and r such that the commuting graph $\Gamma(G)$ is similar to the Erdős–Rényi graph $G_{n,p}$ with $p = n^{-1+\epsilon}$, which is known to have diameter concentrated at $\lceil 1/\epsilon \rceil$ with high probability for small $\epsilon > 0$.

Let $k \geq 2$ be an integer, and $\delta \in \left(0, \frac{1}{2k(k-1)}\right)$ a real number. There is a choice of real number $\delta_1 > 0$ such that the following holds: for each positive integer m, if we set

$$r := \lfloor (1 - \delta_1)m \rfloor, \quad p := 2^{-r}, \quad n := 2^m - 1, \tag{3}$$

then, for all m sufficiently large,

$$1 + \log_n p \in \left(\frac{1}{k} + \delta, \frac{1}{k-1} - \delta\right). \tag{4}$$

The probability that an edge of $\Gamma(G)$ is present is then p, as this is the probability that a uniformly chosen random element of $\mathcal{H}$ is zero. Thus one can hope that its diameter is concentrated around k, as it would be if the states of all edges were independent as in $G_{n,p}$.

Unfortunately, it becomes difficult to translate the known methods of $G_{n,p}$ to our setting due to large amount of dependence between edges, so we were unable to prove this correspondence in full. However, some convincing structural results appear to be amenable to the second moment method.

Proposition 3. *Let $G_{m,k}$ be the group defined above with corresponding parameters m, r and k. Then*

 (i) *As $m \to \infty$, $\mathbb{P}(G' = Z(G) = \mathcal{H}) \to 1$.*
 (ii) *There is some $\delta_3 > 0$, depending on the choices of δ and δ_1, such that, as $m \to \infty$, $\Gamma(G_{m,k})$ almost surely has a connected component of size at least $n - n^{1-\delta_3}$. The diameter of $\Gamma(G_{m,k})$ is at least k w.h.p., but might be infinite if it is not connected.*

So in fact to provide a counterexample to Conjecture 1 it is sufficient to prove that $\Gamma(G_{m,k})$ remains connected for large m and fixed k. We conjecture that even a more precise statement holds.

Conjecture 4. As $m \to \infty$, $\Gamma(G_{m,k})$ is almost surely connected and of diameter k.

ACKNOWLEDGEMENTS. A preliminary version of the current note was presented at a workshop hosted by Scuola Normale Superiore di Pisa, whose hospitality is acknowledged.

References

[1] N. ALON and J. SPENCER, "The Probabilistic Method" (2nd edition), Wiley, 2000.

[2] M. GIUDICI and C. W. PARKER, *There is no upper bound for the diameter of the commuting graph of a finite group*, J. Combin. Theory Ser. A **120**, no. 7 (2013), 1600–1603.

[3] M. GIUDICI and A. POPE, *On bounding the diameter of the commuting graph of a group*, J. Group Theory. **17**, no. 1 (2013), 131–149.

[4] P. HEGARTY and D. ZHELEZOV, *On the Diameters of Commuting Graphs Arising from Random Skew-Symmetric Matrices*, Combin. Probab. Comput. **23**, no. 3 (2014), 449–459.

[5] A. IRANMANESH and A. JAFARZADEH, *On the commuting graph associated with the symmetric and alternating groups*, J. Algebra Appl. **7**, no. 1 (2008), 129–146.

[6] G. L. MORGAN and C. W. PARKER, *The diameter of the commuting graph of a finite group with trivial centre*, J. Algebra **393** (2013), 41–59.

[7] B. H. NEUMANN, *A problem of Paul Erdős on groups*, J. Austral. Math. Soc. Ser. A **21** (1976), 467–472.

CRM Series
Publications by the Ennio De Giorgi Mathematical Research Center Pisa

The Ennio De Giorgi Mathematical Research Center in Pisa, Italy, was established in 2001 and organizes research periods focusing on specific fields of current interest, including pure mathematics as well as applications in the natural and social sciences like physics, biology, finance and economics. The CRM series publishes volumes originating from these research periods, thus advancing particular areas of mathematics and their application to problems in the industrial and technological arena.

Published volumes

1. Matematica, cultura e società 2004 (2005). ISBN 88-7642-158-0
2. Matematica, cultura e società 2005 (2006). ISBN 88-7642-188-2
3. M. GIAQUINTA, D. MUCCI, *Maps into Manifolds and Currents: Area and $W^{1,2}$-, $W^{1/2}$-, BV-Energies*, 2006. ISBN 88-7642-200-5
4. U. ZANNIER (editor), *Diophantine Geometry*. Proceedings, 2005 (2007). ISBN 978-88-7642-206-5
5. G. MÉTIVIER, *Para-Differential Calculus and Applications to the Cauchy Problem for Nonlinear Systems*, 2008. ISBN 978-88-7642-329-1
6. F. GUERRA, N. ROBOTTI, *Ettore Majorana. Aspects of his Scientific and Academic Activity*, 2008. ISBN 978-88-7642-331-4
7. Y. CENSOR, M. JIANG, A. K. LOUISR (editors), *Mathematical Methods in Biomedical Imaging and Intensity-Modulated Radiation Therapy (IMRT)*, 2008. ISBN 978-88-7642-314-7
8. M. ERICSSON, S. MONTANGERO (editors), *Quantum Information and Many Body Quantum systems*. Proceedings, 2007 (2008). ISBN 978-88-7642-307-9
9. M. NOVAGA, G. ORLANDI (editors), *Singularities in Nonlinear Evolution Phenomena and Applications*. Proceedings, 2008 (2009). ISBN 978-88-7642-343-7
   – Matematica, cultura e società 2006 (2009). ISBN 88-7642-315-4
10. H. HOSNI, F. MONTAGNA (editors), *Probability, Uncertainty and Rationality*, 2010. ISBN 978-88-7642-347-5

11. L. AMBROSIO (editor), *Optimal Transportation, Geometry and Functional Inequalities*, 2010. ISBN 978-88-7642-373-4

12*. O. COSTIN, F. FAUVET, F. MENOUS, D. SAUZIN (editors), *Asymptotics in Dynamics, Geometry and PDEs; Generalized Borel Summation*, vol. I, 2011. ISBN 978-88-7642-374-1, e-ISBN 978-88-7642-379-6

12**. O. COSTIN, F. FAUVET, F. MENOUS, D. SAUZIN (editors), *Asymptotics in Dynamics, Geometry and PDEs; Generalized Borel Summation*, vol. II, 2011. ISBN 978-88-7642-376-5, e-ISBN 978-88-7642-377-2

13. G. MINGIONE (editor), *Topics in Modern Regularity Theory*, 2011.
ISBN 978-88-7642-426-7, e-ISBN 978-88-7642-427-4

– Matematica, cultura e società 2007-2008 (2012).
ISBN 978-88-7642-382-6

14. A. BJORNER, F. COHEN, C. DE CONCINI, C. PROCESI, M. SALVETTI (editors), *Configuration Spaces*, Geometry, Combinatorics and Topology, 2012. ISBN 978-88-7642-430-4, e-ISBN 978-88-7642-431-1

15 A. CHAMBOLLE, M. NOVAGA E. VALDINOCI (editors), *Geometric Partial Differential Equations*, 2013.
ISBN 978-88-7642-343-7, e-ISBN 978-88-7642-473-1

16 J. NEŠETŘIL AND M. PELLEGRINI (editors), *The Seventh European Conference on Combinatorics, Graph Theory and Applications*, EuroComb 2013. ISBN 978-88-7642-524-0, e-ISBN 978-88-7642-525-7

17 L. AMBROSIO (editor), *Geometric Measure Theory and Real Analysis*, 2014. ISBN 978-88-7642-522-6 , e-ISBN 978-88-7642-523-3

18 J. MATOUŠEK, J. NEŠETŘIL, M. PELLEGRINI (editors), *Geometry, Structure and Randomness in Combinatorics*, 2015.
ISBN 978-88-7642-524-0, e-ISBN 978-88-7642-525-7

Volumes published earlier

Dynamical Systems. Proceedings, 2002 (2003)
    Part I: *Hamiltonian Systems and Celestial Mechanics*.
ISBN 978-88-7642-259-1
    Part II: *Topological, Geometrical and Ergodic Properties of Dynamics*.
ISBN 978-88-7642-260-1

Matematica, cultura e società 2003 (2004). ISBN 88-7642-129-7

Ricordando Franco Conti, 2004. ISBN 88-7642-137-8

N.V. KRYLOV, *Probabilistic Methods of Investigating Interior Smoothness of Harmonic Functions Associated with Degenerate Elliptic Operators*, 2004. ISBN 978-88-7642-261-1

Phase Space Analysis of Partial Differential Equations. Proceedings, vol. I, 2004 (2005). ISBN 978-88-7642-263-1

Phase Space Analysis of Partial Differential Equations. Proceedings, vol. II, 2004 (2005). ISBN 978-88-7642-263-1

Fotocomposizione "CompoMat" Loc. Braccone, 02040 Configni (RI) Italia
Finito di stampare nel mese di gennaio 2015
dalla CSR, Via di Pietralata 157, 00158 Roma